EcoProduction

Environmental Issues in Logistics and Manufacturing

Series Editor

Paulina Golinska, Poznan, Poland

For further volumes:
http://www.springer.com/series/10152

About the Series

The EcoProduction Series is a forum for presenting emerging environmental issues in Logistics and Manufacturing. Its main objective is a multidisciplinary approach to link the scientific activities in various manufacturing and logistics fields with the sustainability research. It encompasses topical monographs and selected conference proceedings, authored or edited by leading experts as well as by promising young scientists. The Series aims to provide the impulse for new ideas by reporting on the state-of-the-art and motivating for the future development of sustainable manufacturing systems, environmentally conscious operations management and reverse or closed loop logistics.

It aims to bring together academic, industry and government personnel from various countries to present and discuss the challenges for implementation of sustainable policy in the field of production and logistics.

Marina G. Erechtchoukova
Peter A. Khaiter · Paulina Golinska
Editors

Sustainability Appraisal: Quantitative Methods and Mathematical Techniques for Environmental Performance Evaluation

Springer

Editors
Marina G. Erechtchoukova
Peter A. Khaiter
Faculty of Liberal Arts
and Professional Studies
School of Information Technology
York University
Toronto, ON
Canada

Paulina Golinska
Poznan University of Technology
Poznan
Poland

ISSN 2193-4614 ISSN 2193-4622 (electronic)
ISBN 978-3-642-32080-4 ISBN 978-3-642-32081-1 (eBook)
DOI 10.1007/978-3-642-32081-1
Springer Heidelberg New York Dordrecht London

Library of Congress Control Number: 2012956163

Printed on acid-free paper

Springer is part of Springer Science+Business Media (www.springer.com)

Preface

One of the important issues in developing sustainable management strategies is the assessment of the sustainability of business operations and an organization's overall environmental performance. The concept of sustainability is multi-faceted and interdisciplinary by nature. To evaluate sustainability in quantitative terms, a large number of indicators, processes, and their interrelations must be taken into account and both present and future values of various indicators must be generated. Traditionally, sustainability is assessed based on a single aspect and using isolated indicators. The urgency to provide an integrated sustainability assessment is well understood, while quantitative methods for such assessment are yet to be developed. It is necessary to identify and bridge gaps between quantitative indicators of different aspects of sustainability for an integrated sustainability appraisal.

This volume presents original research papers on the state-of-the-art in sustainability appraisal, including the development and application of sustainability indices, quantitative methods, multi-criteria optimization models, and frameworks for evaluation of an organization's environmental performance, as well as eco-efficiency approaches leading to business process re-engineering and modern trends in environmental reporting. By its scope, the book is intended for a broad audience from the academia and the industry and can be of interest for environmental researchers, business managers and process analysts, information management professionals and environmental decision makers who will find valuable sources of information for their work-related activities.

The book showcases contributions of geographically dispersed authors from Europe, North America, and Asia. It is a clear indication of a growing interest in green economy and international collaboration on the issues of sustainable development. The chapters in the volume explore international approaches to sustainability assessment as well as their country-specific applications. The high scientific quality of the chapters was assured by a rigorous double-blind review process implemented by the leading researchers in the field from Australia, Denmark, Canada, Germany, Italy, New Zealand, Poland, Spain, United Kingdom, and USA.

The volume is published in the EcoProduction series and, as one of its milestones, aims to disseminate new ideas and motivate future developments for integrating sustainability concepts into product systems.

Our project would not be successful without its key participants—Authors and Reviewers. We would like to thank all researches who responded to the call for chapters and submitted manuscripts to this volume for their interest in the project. Although not all of the received papers appear in this book, the efforts spent and the work done for this project are very much appreciated. We would like to thank all authors of the chapters in this book for their hard work on manuscripts under tight deadlines and high quality of the contributions presented.

We are grateful to our reviewers whose names are not listed in the volume due to the confidentiality of the process. Their voluntary service and insightful comments helped the authors to improve the quality of the manuscripts as well as to make editorial decisions on each chapter.

We would like to thank Dr. M. Singer, Dean, Faculty of Liberal Arts and Professional Studies, York University, Canada for his support of this project. The allocated time was much needed to complete the book.

We would like to express our appreciation to Mr. Thambidurai Solaimuthu, Springer Project Coordinator (Books) for his constructive guidance and friendly support of the project from the beginning throughout all of its stages.

Marina G. Erechtchoukova
Peter A. Khaiter
Paulina Golinska

Contents

Dimensions of Sustainability Appraisal

Introduction

Marina G. Erechtchoukova and Peter A. Khaiter

Abstract Quantitative assessment of sustainability becomes an important issue for selecting technological and managerial alternatives in all areas of human activities including industry and policy making. The main frameworks for sustainability appraisal and their evolution are briefly described in the chapter. These frameworks govern the development of sustainability indicators with various levels of granularity. Sustainability assessment is conducted as a comparison and analysis of values of current and future welfare outcomes, which make the application of models and mathematical tools unavoidable and explain the necessity to use quantitative indicators of sustainability. The chapter presents an overview of the book.

Keywords Sustainability indicators · Environmental performance evaluation · Eco-efficiency

1 Eco-efficiency in View of Sustainability

The term 'eco-efficiency' was introduced by the World Business Council for Sustainable Development in 1992. Eco-efficiency was formulated as a way to convey sustainability to the entire life-cycle of products and as one of "the new

M. G. Erechtchoukova (✉) · P. A. Khaiter
School of Information Technology, Faculty of Liberal Arts and Professional Studies, York University, Toronto, Canada
e-mail: marina@yorku.ca

P. A. Khaiter
e-mail: pkhaiter@yorku.ca

M. G. Erechtchoukova et al. (eds.), *Sustainability Appraisal: Quantitative Methods and Mathematical Techniques for Environmental Performance Evaluation*, EcoProduction, DOI: 10.1007/978-3-642-32081-1_1,

forms of cooperation between government, business, and society" (Schmidheiny 1992). Since then, it has been refined and elaborated into a management strategy incorporating three main objectives: increasing the value of a product or a service, optimizing all resources used towards that increase and reducing the associated impact on the environment. Quantitative eco-efficiency assessment of product systems becomes an important process for choosing correct technological and managerial alternatives. This fact is reflected in the recent ISO 14045 standard. The standard outlines several main components of the eco-efficiency assessment including environmental assessment, the product-system-value assessment, the quantification of eco-efficiency, and reporting (ISO 14045 2012). Similarly to other ISO standards, it does not supply a rigid set of formulae or methods for evaluation, but it rather presents a framework which is expected to be filled with quantitative techniques specific to business activities. The scope of the standard clearly indicates that sustainability appraisal is an integral part of the process of developing eco-efficient strategies, alongside environmental assessment and reporting. All these activities are examined in the chapters of this book.

Recognition that changes in the state of natural environments have complex causes and even more complex consequences motivates the development of a structured approach to managing environmental issues and their interrelations with political and economic processes. By its nature, sustainability appraisal is a multidisciplinary analysis of environmental problems at different scales with respect to their effects on the economy and society. Jeffry (2005) classified the main aspects of integration as (1) integration across disciplinary, professional or cultural perspectives; (2) investigation of cause-effect chains; (3) integration across spatial and temporal scales; (4) legal and social constraints; and (5) "the description-diagnosis-prescription" process. Each aspect implies specific analytical techniques that may vary from case to case.

2 Assessment and Reporting Frameworks Promoting Sustainability

The systems approach considers the environment as a component which affects and receives impacts from other components of a larger system. Economy and society are the two other pillars closely interacting with the environment. In addition, the society applies an institutional component to manage interactions between all the pillars and activities within them (Rotmans and van Asselt 2001; Thornton et al. 2006). This representation allows for distinguishing between the three main aspects of possible consequences of a policy or an undertaking in terms of environmental, economic and social impacts. The aspects represent the main themes of the environmental assessment.

A refined interaction of the environment, economy and society was described by the Pressure-State-Response framework adopted by the EU Organization for

Economic Co-operation and Development in early 1990s' (e.g., OECD 1997). This framework was elaborated to reflect a more comprehensive approach to the analysis of environmental problems. As a result, a detailed framework, the drivers-pressures-states and trends-impacts-responses (DPSIR) was proposed (Gabrielsen and Bosch 2003). It identifies relationships between the main components, important for integrated assessment of anthropogenic impact on the environment. The conceptual scheme of the framework considers two main actors: the environment and the society. 'Drivers' (i.e., economic and/or societal processes) create 'pressure' on the environment. The environment alters its 'state' under this pressure and inevitably changes the services it provides to the society. The latter constitutes an 'impact' on the society. 'Responses' to the impact generated by the society may represent either adaptation to environmental changes or reduction of the pressure. The DPSIR framework has been modified lately by adding new findings in environmental assessment and also by extending regional specifics in the environmental reporting.

Reflecting the multidimensional nature of human-environment interactions, a building block approach has been introduced (UNEP 2009). The building blocks can be used in sets with different numbers and sequences of blocks in order to tailor the assessment to different types of policymaking and planning processes. All blocks are classified into three groups describing the process itself, the policy institutional context and the analytical context. The first group provides blocks to design the assessment process, including communication links between different stages in a way that fits the policy-making procedures. The policy institutional context ensures stakeholder engagement, and investigation and the improvement of institutional context, and evaluation of the effectiveness of the assessment process. The analytical context determines actual steps undertaken for the assessment.

These frameworks have been explicitly declared as conceptual models for the integrated environmental assessment mainstreaming sustainability. However, to be applied, either framework has to be filled with issue-specific indicators. These indicators depend on the problem domain as well as on the temporal and spatial scales of an investigated case study. At the same time, these indicators must be consistent with commonly recognized sustainability measures recommended for environmental reporting.

3 Sustainability Indicators

An assessment framework requires the selection of indicators relevant to the goal and evaluating them as a means for comparison of a strategy or a process with its alternatives. A commonly recognized necessity to describe sustainable development by a set of quantitative indicators measuring sustainability at the national and international levels has been expressed in Agenda 21 (UN 1992). The conceptual frameworks for sustainability indicators help to answer the key questions: (a) what to measure? (b) what to expect from the measurements? and (c) what kind of indicators to apply? The first set of sustainability indicators has been developed based on the Drivers-States-Responses framework, where 'drivers' represent activities affecting sustainable development, 'states' reflect the current status, and 'responses' describe societal actions directed towards sustainable development.

Since the framework was introduced, the set of sustainability indicators has been significantly revised. As a result, the original 134 indicators have been reduced to 58 and further modified to 96 indicators of sustainable development which contain 50 core indicators according to evolving perceptions on the assessment and reporting process and its DPSIR framework, existing practice and the feasibility of some indicators. Core indicators are common for most countries. These indicators are irreducible since they supply information unavailable through other indicators from the set. Finally, these indicators are economically feasible as they can be evaluated on the data which are either available in the most of the countries or can be obtained within reasonable time and expense (UN 2007).

The DPSIR framework supports the assessment of environmental sustainability mainly from the environmental perspective, i.e., reflects the extent, to which the anthropogenic impacts are adverse and how they can be mitigated. Economic perspectives suggest indicators that express interactions between economic processes and natural environment in monetary terms. The System of National Accounts quantifies economic activities related to production, consumption, investment and foreign trade. An introduction of similar indicators to express natural capital and its changes under anthropogenic impacts seems promising for the integration of the assessment. However, monetary estimates of many environmental impacts do not exist at present due to disagreement on impact appraisal and availability of relevant data. Rather, the System of Integrated Environmental and Economic Accounts extends existing national accounts to describe the environment in physical terms (UN 2003).

Indicators for the Social pillar have been developed as measures of the quality of human life. The System of Social and Demographic Statistics attempted to describe stocks and flows of individuals and their involvements in economic and social processes in an accounting form reflecting "life sequences, time budgets and cost-benefit distributions" (Bartelmus 2008). Unfortunately, no agreement has been reached on the concept of the quality of life and its quantification. Unified theories or data systems for environment-population interactions are less developed than those describing economy-environment interplay. As a result, the assessment must take into account important demographic and socio-economic characteristics such as migration, fertility, marital status, education, labour force status, occupation, industry and health.

To evaluate the sustainability of an undertaking, it is necessary to obtain and analyze values of current and future welfare outcomes. The latter makes an application of models and mathematical tools unavoidable and justifies the necessity to use quantitative indicators of sustainability. Modelling is charged with evaluation of indicators of environmental, economic and social impacts of an activity, an institutional change or an event based on available data and transforming these data into information for decision makers. Each modelling exercise is case-specific. However, the common challenge of any modelling exercise is the necessity to combine indicators of the states of three main pillars into a single framework and present the results of quantitative assessment to decision makers in a clear and persuasive form.

4 Structure of the Book

The book presents original research chapters on quantitative methods for sustainability appraisal at the international, national and local levels, on application of quantitative methods for developing environmentally friendly business strategies. It can be viewed as a reference source for environmental researchers, business managers and process analysts, information management professionals and environmental decision makers who will find valuable information for their professional activities.

The book is subdivided into three main parts. The first part is focused on sustainability indicators. The chapters in this part explore a methodology for quantitative comparison of environmental performance between countries based on sustainability indices, approaches to measuring and evaluating business sustainability on the basis of corporate index of sustainability performance. The last chapter in this section concerns the development and application of sustainability indicators in agriculture.

The second part explores trends in environmental reporting, key indicators of an organization's environmental performance in different industrial areas, and approaches to eco-production.

The third part discusses ways of improving eco-efficiency by applying techniques of multi-criteria decision making including operations research methods and the analytic hierarchy method. The chapters in this part present case studies from the energy sector, public transportation systems and product systems.

References

Bartelmus P (2008) Quantitative Eco-nomics. How sustainable are our economies? Springer Science + Business media

Gabrielsen P, Bosch P (2003) Environmental indicators: typology and use in reporting. EEA internal working paper. European environment agency, August 2003

ISO 14045:2012 (2012) Environmental management—Eco-efficiency assessment of product systems—Principles, requirements and guidelines. http://www.iso.org/iso/catalogue_detail?csnumber=43262 Accessed 22.11.2012

Jeffry P (2005) Coming home; a personal re-evaluation of the aspirations and utility of integrated assessment. Integr Assess J 5(1):27–40

OECD (1997) Environmental performance reviews—a practical introduction" OCDE/GD(97)35 Paris

Rotmans J, van Asselt MBA (2001) Uncertainty management in integrated assessment modeling: towards a pluralistic approach. Environ Monit Assess 69(2):101–130

Schmidheiny S (1992) Changing course: a global business perspective on development and the environment. MIT Press, Cambridge

Thornton KW, Laurin C, Shortle J, Fisher A, Sobrino J, Stewart M (2006) A framework and guidelines for moving toward sustainable water resources management. In: Proceedings of the water environment federation, WEFTEC2006, pp. 2762–27777

UNEP (2009) Integrated assessment: mainstreaming sustainability into policymaking. A guidance manual. http://www.unep.ch/etb/publications/AIguidance2009/UNEPIAfinal.pdf. Accessed 10.12.10

United Nations (1992) The united nations programme of actions from Rio. Earth summit. http://www.un.org/esa/dsd/agenda21/index.shtml. Accessed 10.12.12

United Nations (2003) Integrated environmental and economic accounting. Handbook of national accounting. http://unstats.un.org/unsd/envAccounting/seea2003.pdf. Accessed 30.04.2011

United Nations (2007) Indicators of sustainable development: guidelines and methodologies. 3d ed., New York. http://www.un.org/esa/sustdev/natlinfo/indicators/guidelines.pdf. Accessed 10.12.10

Part I
Sustainability Indicators

The FEEM Sustainability Index: An Integrated Tool for Sustainability Assessment

Carlo Carraro, Lorenza Campagnolo, Fabio Eboli, Silvio Giove, Elisa Lanzi, Ramiro Parrado, Mehmet Pinar and Elisa Portale

Abstract The FEEM Sustainability Index (FEEM SI) proposes an integrated methodological approach to quantitatively assess sustainability performance across countries and over time. Three are the main features of this approach: (1) the index considers sustainability based on economic, environmental and social indicators simultaneously; (2) the framework used to compute the indicators, i.e. a Computable General Equilibrium (CGE) model, allows to generate projections on

---

C. Carraro · L. Campagnolo · F. Eboli · E. Lanzi ·
R. Parrado (✉) · M. Pinar · E. Portale
Fondazione Eni Enrico Mattei (FEEM), Isola di San Giorgio Maggiore, 30124 Venice, Italy
e-mail: ramiro.parrado@feem.it

C. Carraro
e-mail: carlo.carraro@feem.it

L. Campagnolo
e-mail: lorenza.campagnolo@feem.it

F. Eboli
e-mail: fabio.eboli@feem.it

E. Lanzi
e-mail: elisa.lanzi@feem.it

M. Pinar
e-mail: mehmet.pinar@feem.it

E. Portale
e-mail: feemsi@feem.it

C. Carraro · F. Eboli · R. Parrado · M. Pinar
Euro Mediterranean Centre on Climate Change (CMCC), Venice, Italy

C. Carraro · S. Giove
Ca' Foscari University of Venice, Venice, Italy
e-mail: sgiove@unive.it

M. G. Erechtchoukova et al. (eds.), *Sustainability Appraisal: Quantitative Methods and Mathematical Techniques for Environmental Performance Evaluation*, EcoProduction, DOI: 10.1007/978-3-642-32081-1_2,

the future evolution of sustainability; and (3) the methodology used for the normalisation and aggregation of the indicators delivers a unique and comprehensive measure of sustainability. These features along with the multi-regional nature of the CGE model consent to perform policy evaluations and sustainability assessments for different countries or regions in the world. This chapter offers a methodological overview of the FEEM SI approach. To illustrate the potential of the methodology for the measurement of sustainability, the chapter also illustrates results from a climate policy scenario. In the mitigation scenario considered Annex I and Non-Annex I countries taking action towards climate change achieve the lower end of the pledges proposed at the 15th UNFCCC Conference of the Parties in Copenhagen. For countries putting into practice the policy, the environmental sphere more than offsets the related costs (economic pillar), leading to an overall improvement in sustainability. At world level, the outcome is positive even though carbon leakage in countries that are not acting reduces the effectiveness of the policy and the sustainability performance.

Keywords Sustainability · Composite indicators · Computable general equilibrium model · Climate policy

1 Introduction

Sustainable development is a paradigm that considers several aspects of growth in a comprehensive framework. The Bruntland Report (WCED 1987) defines it as "development that meets the needs of the present without compromising the ability of future generations to meet their own needs". Two are the main concepts comprised in this paradigm: (1) the simultaneous achievement of economic, social and environmental sustainability, and (2) the intra/intergenerational equity.

The most recent evolution of the sustainability debate refers to the analysis developed by Stiglitz, Sen and Fitoussi (Commission on the Measurement of Economic Performance and Social Progress 2009). This tries to define more concretely the concept of sustainable development and clarify the methodological approaches in this field. The "Rio + 20" conference (June 2012) assessed the main achievements in sustainable development in the last 20 years, providing further guidelines with main focus on the green economy and the effective integration of sustainable development within all levels of institutional governance. The outcome of the conference underlined the importance of tracking sustainability, as suggested by the statement that "progress towards the achievement of the goals needs to be assessed and accompanied by targets and indicators, while taking into account different national circumstances, capacities and levels of development" (UN 2012).

A valid tool to measure sustainability is a set of indicators (Parris and Kates 2003; Singh et al. 2009). Thanks to their synthetic properties, indicators are widely used in policymaking and public communication. Further, substantial efforts have been devoted to create lists of indicators that address the concept of sustainable development in a comprehensive way (United Nations' Commission on Sustainable Development- UNCSD; European Union's Sustainable Development Strategy—EU SDS; World Bank's World Development Indicators—WDI). Research has focused mostly on expanding the sustainability dimensions considered or on the selection of appropriate indicators. There have also been a few attempts at aggregating indicators to indices, which are generally focused on a specific area of sustainability. Many aggregate measures are nowadays used in policymaking and assessments. Examples are: (1) the HDI—Human Development Index (UNDP 1990), (2) GS—Genuine Savings (Yusuf et al. 1989), (3) the ISEW—Index of Sustainable Economic Welfare (Daly and Cobb 1989), and (4) the EPI—Environmental Performance Index (Yale and Columbia Universities 2010). These aggregate indices generally focus on one precise aspect of sustainability.

The indicators' aggregation procedure is a controversial issue. However, an index built with a transparent aggregation methodology and complementary to its single components can be very useful for summarising a wide range of information. Such an index facilitates policy design, assessment and implementation, and allows to explore the trade-offs and relationships among indicators.

In this context, the Fondazione Eni Enrico Mattei (FEEM) has been working on developing a new tool for sustainability assessment—the FEEM Sustainability Index (FEEM SI)—since 2006.[1] A first version was released in 2009 while the updated structure for its second release (2011) is presented in this chapter. The index summarises and merges information derived by a selection of relevant sustainability indicators offering a more comprehensive account of sustainability.

The FEEM SI is an aggregate index composed of a set of indicators that captures the main elements of sustainable development (socio-economic and environmental components). The index uses a specific aggregation methodology that considers the interactions among indicators by relying on subjective experts' evaluations. As it is built in a recursive-dynamic Computable General Equilibrium (CGE) model, the FEEM SI can be used to analyse and compare sustainability across different policy scenarios. This allows including in the analysis the intertemporal aspects of sustainability. While the nature of the macroeconomic model implies some drawbacks (e.g., the absence of indicators disconnected from economic activity), the modelling framework provides a coherent context for calculating indicators with comparability across countries, time and alternative scenarios.

To illustrate the potential of the methodology to measure sustainability, this chapter also illustrates results from a climate policy scenario. In the mitigation scenario considered Annex I and Non-Annex I countries taking action towards climate change achieve the low pledges proposed at the 15th UNFCCC Conference of the Parties in Copenhagen (December 2009). The results show that, for countries putting into practice the policy, the environmental sphere more than

[1] The complete overview on methodology and results is available at: www.feemsi.org. See also Carraro et al. (2012).

offsets the related costs (economic pillar), leading to an overall improvement in sustainability performance. At world level, the outcome is positive even though carbon leakage in countries that are not acting reduces the effectiveness of the policy and the sustainability outcome.

The structure of the chapter is as follows. Section 2 describes the composition of the FEEM SI and its indicators. Section 3 presents the CGE approach and the necessary extensions of both the database and the model to compute the indicators. Section 4 illustrates the normalisation and aggregation methodology. Section 5 presents the main results for a baseline scenario while Sect. 6 considers the effects of a climate policy on sustainability. Section 7 concludes.

2 The FEEM SI Structure

The list of indicators included in the FEEM SI has been determined after a thorough analysis of the sustainable development literature. The selection process has been further refined to consider only indicators manageable in the framework of the macroeconomic model used for scenario building. The world coverage requires data availability for the entire world at country or macro-region scale. The specific methodology applied to define future sustainability limited the choice to indicators that can be directly linked to economic measures present in the model.

Figure 1 illustrates the structure of the FEEM SI and includes all indicators selected for the index construction. Along with the wide definition of sustainability, the structure of the tree considers its three main pillars: economic, social and environmental. For each of these dimensions, the FEEM SI tree covers the main areas of sustainability assessment: economic growth drivers, GDP per capita, economic exposure; population density, well-being, social vulnerability; energy, air quality, and natural endowments.

Table 6 of Annex I summarises the indicators selection and describes the indicators, including their affiliation to a particular area of sustainability, definition, implementation in the model and relevant references to the literature.

3 Modeling Framework

Processing sustainability indicators within the framework of a CGE model has a number of advantages. One of the main features of CGE models is to consider the interactions existing within and across productive systems in a consistent framework. This contributes to increase the comparability of the different indicators. Further, as argued by Böhringer and Löschel (2006), CGE models also allow performing a trade-off analysis among different components of sustainability. This feature is especially useful in analysing the effects of a policy implementation. An intervention in one dimension of sustainability in a specific country will influence

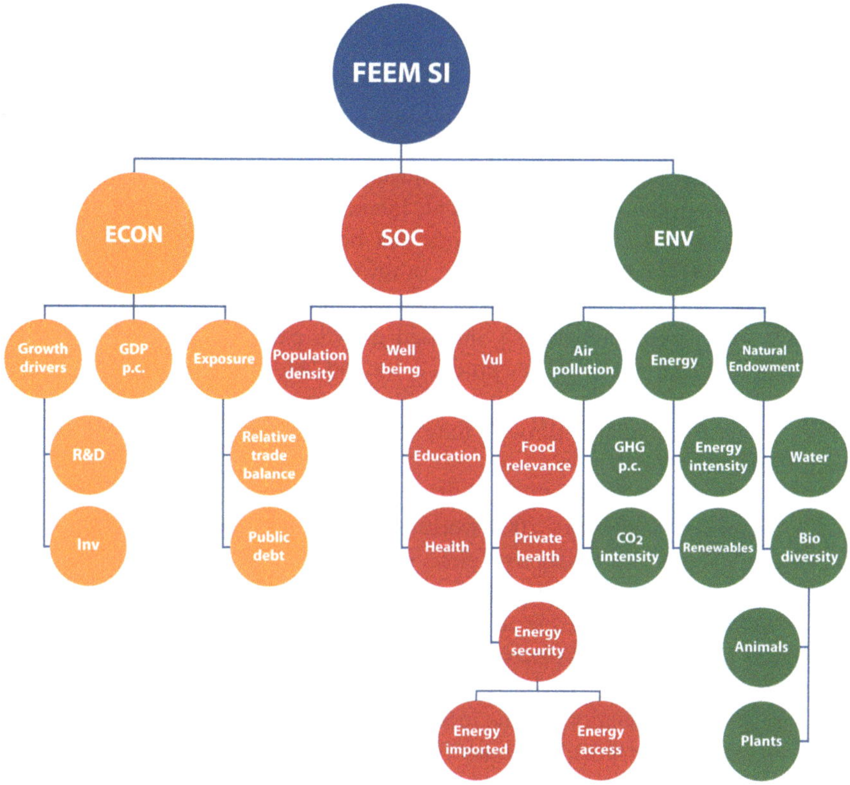

Fig. 1 FEEM SI 2011 indicators' tree

other aspects of sustainability in that country as well as in other countries. Finally, when using a dynamic CGE model, it is also possible to make projections of the indicators and thus perform a scenario analysis of future sustainability under different policy proposals.

The main difficulty in using a quantitative economic model is to link environmental and social indicators to economic variables computed in the model. This reflects a limited flexibility in defining a full set of indicators. Some of these indicators, which are not directly connected to specific economic activities, may play a role in assessing sustainability but can hardly be modelled to depict their future evolution.

The CGE model used—ICES-SI[2]—is an ideal framework for the construction of a policy-oriented sustainability index. The model allows to compute indicators related to different productive sectors and calculating the index for each region in the world (either at national or macro-regional level). Furthermore, its dynamic

[2] A detailed description of the model tailored to be used for sustainability indicators is in the FEEM SI Methodological report (FEEM 2011) and Carraro et al. (2012).

framework generates scenarios that can be used to calculate the index in the future under different policy assumptions.

Within the CGE framework, industries are modelled as a representative cost-minimizing firm with nested production functions in which primary factors and intermediates are combined to produce the final output. A representative household in each region receives income, defined as the service value of the national primary factors (natural resources, land, labour, and capital). Demand for production factors and consumption goods can be satisfied either by domestic or foreign producers that are not perfectly substitutable (Hertel 1997). The dynamic of the model is driven by two sources: one exogenous and the other endogenous. The first stems from exogenously imposed growth paths for some key variables (population, labour stock, labour productivity and land productivity). The second concerns the endogenous process of capital accumulation, according to which capital stock is cumulated through time taking into account endogenous investment decisions.

ICES-SI is based on the GTAP 7 database (Narayanan and Walmsley 2008), which presents a snapshot of 2004 world economic flows. The world economy is divided in 40 countries or macro-regions in which countries are at a similar stage of development or have similar characteristics (see Table 7 in Annex I). Within each country/macro-region, the economy is represented by 20 sectors (see Table 8 in Annex I). In order to perform the analysis on future sustainability trends throughout the world, the ICES-SI sectoral details have been enhanced by adding new variables and equations to the model. This allows increasing its flexibility in capturing as many as possible dimensions of sustainable development.

A number of indicators are sector-specific in the sense that they refer to their share of expenditure or production over GDP (i.e. Health or Education expenditure are used as indicator for the social pillar) or output of a subset of productive sectors (i.e. Renewables demand over total energy demand). Some sustainability indicators focus on sectors that are not represented in the original GTAP 7 database. In order to increase the informative purpose of the Index, the original database has been modified to increase the sectors specification. Research and Development (R&D), Education, Private and Public Health, and Renewables have been included in the model using data on trade flows, production and consumption from different sources (Table 1).[3] These new sectors have an endogenous evolution coherent with the exogenous assumptions on primary factors' productivity.

Other indicators focus on variables that are not part of the ICES-SI model, namely use of water, biodiversity, access to electricity and inhabitable land. The above variables have been linked to the model with additional equations that allow simulating their future behaviour coherently with the endogenous path of ICES-SI. Table 2 reports the way in which the new indicators are linked to the model and main sources for data collection with relation to the base year.

---

[3] These sectors were extracted from more aggregate sectors by using the SplitCom facility (Horridge 2008) and constructed using data from relevant sources.

Table 1 Additional sectors for FEEM SI

New sector	Original GTAP7 sector	Main reference sources
R&D	"Other business services"	World Bank (2010a)
Education/private health/ public health	"Other generative services"	World Bank (2010b), WHO (2010)
Renewables	"Electricity"	IEA (2005), EC (2008), Ragwitz et al. (2007), GTZ (2009), IEA country profiles, REN21 (2011) renewable energy policy network for the 21st century (www.ren21.net)

Table 2 Additional sectors for FEEM SI

New indicator	Model variable link	Main reference sources
Water	Water services demand by agriculture, industry and households	FAO's aquastat database
Biodiversity	CO_2 emissions	World conservation union IUCN (2010), Thomas et al. (2004)
Electricity access	GDP per capita	IEA (2010), World Bank (2010b)
Inhabitable land	Population	FAO (2011); FAO and IIASA (2000)

The Water sector in GTAP7 refers to infrastructure whose services by agriculture, industry and households were used to consider the exploitation of water, keeping constant the available total renewable water resources in each country. Biodiversity has been assumed to decline with increases in carbon dioxide emissions. Reducing GDP per capita gap with respect to developed countries allows reproducing an increase in access to electricity in developing countries. Finally, growing population raises the pressure over the inhabitable land.

The physical energy flows underlying the database (production, consumption and trade of energy) and the Kyoto GHG emissions (CO_2, CH_4, N_2O, PFCs, HFCs, SF_6) (Lee 2008; Rose and Lee 2008), are included to consider GHG per capita, energy intensity and CO_2 intensity. They evolve coherently with economic flows.

4 Normalisation and Aggregation Procedure

The output of the ICES-SI model provides the initial values for the indicators that are then normalised and aggregated. The idea of having comparable indicators and one index to assess the overall level of sustainability, across countries and time, requires two main steps.

To begin with, it is necessary to express all indicators, characterised by different measure units, in a common measurement scale. According to the OECD's *Handbook on constructing composite indicators* (2008), "normalisation is required prior to any data aggregation as the indicators in a data set often have different measurement units". Several normalisation techniques exist in literature. The FEEM SI normalisation method uses a mixed strategy. First, a *re-scaling* procedure is applied to all indicators to obtain values in the range [0, 1], where 0 defines extremely unsustainable and 1 fully sustainable performance. Second, a step-wise *benchmarking* function is defined for each indicator in order to consider intermediate levels of performance.

The use of a benchmarking procedure is appropriate in the case of indicators for which a policy target or a minimum/maximum threshold exists for the extremely unsustainable or fully sustainable levels for the indicators respectively. This method allows comparison through time and across countries, whilst supplying a policy-based normalisation, which is particularly suitable for the construction of the FEEM Sustainability Index. Rather than subtracting mean and dividing each indicator by its standard deviation, we supply a benchmark for sustainable targets. Therefore, our index aims for absolute sustainable level of each indicator and country rather than their relative positions to the highest or lowest levels of each indicator. Since the purpose of creating a sustainability index is not only to identify best and worst practices, but also to give an appraisal of the relative distance to the sustainable target, the FEEM SI indicators are normalised according to a benchmark function, which passes through five reference levels.[4]

To avoid the discontinuity of a step function, each level has been "linearised" taking the mean values of two subsequent intervals and interpolating them, thereby creating a continuous step function (Fig. 2). The intervals are defined considering both relevant literature and official statistics to derive the most appropriate benchmarks for each indicator.

When all indicators are expressed in the [0,1] range through normalization, the next step is the aggregation of all indicators in one general index. This is a three-stage procedure considering: (1) evaluation elicitation, (2) aggregation of single preferences in a representative profile of weights, and (3) index computation combining weights and normalised indicators.

The first stage is the definition of weights to be associated to each indicator. To this purpose, an experts' elicitation with an "ad hoc" questionnaire is performed. The questionnaire is prepared in such a way that experts were asked to evaluate all possible scenarios of the indicators being at their best or worst levels, i.e. all the combinations of BEST and WORST values, as well as how they would evaluate intermediate conditions. Firstly, they are asked to evaluate all possible conditions when only one sustainability indicator is completely sustainable (i.e., best), but the remaining ones are completely unsustainable (i.e., worst). Secondly, they are asked

[4] The complete description of the normalization and benchmarking procedure as well as the benchmark selection is in Chap. 3 of the FEEM SI Methodological report (FEEM 2011).

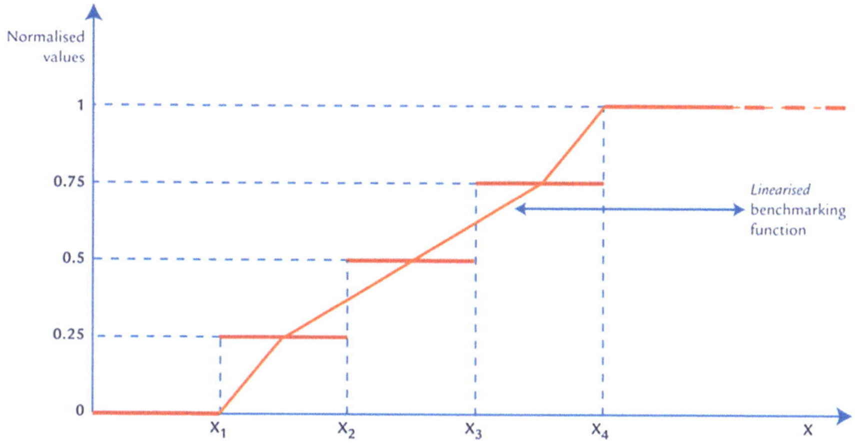

Fig. 2 The benchmarking function

to evaluate all possible combinations when two sustainability indicators are completely sustainable (i.e., best) and the remaining one is completely unsustainable (i.e., worst). Similar types of questions allow evaluating the indicators located under each node in the decision tree.

In the second stage, a non-linear aggregation methodology is applied to aggregate divergences in respondents and to compute a consensus measure. This allows to derive a 'representative' weight assigned to each sustainability indicator and tree's node, relying upon the *metric distance* measure (i.e., if the evaluation of an expert is in agreement with other experts, then this expert's valuation gets higher weight. Thus, if an expert's valuation of sustainability indicators is extremely different from other experts, a relatively lower weight is assigned to this type of expert valuation).

The third stage concerns the aggregation of indicators, combining normalised indicators' values and their weights created in the previous step. The aggregated Sustainability Index is constructed through a non-linear aggregation methodology, the Choquet integral, which accounts for the possible interactions among sustainability indicators (see Murofushi and Soneda 1993; Murofushi et al. 1994; Grabisch 1995, 1996; Marichal and Roubens 2000; Grabisch et al. 2003 for the detailed Choquet integral aggregation procedure and its characteristics).[5] For the aggregation, the decision tree should be read from bottom (leaves) to top (final node) and the tree respects the three main pillar structure which is quite standard in most sustainability studies (see e.g., UN CSD 2005; Global Reporting Initiative framework, GRI 2006, 2010; Krajnc and Glavic 2005), with the final node producing the aggregate index. Finally, economic, social and environmental

[5] See Meyer and Ponthière (2011) for a recent application of the Choquet integral to construct a ranking of multiattribute hypothetical societies by eliciting individual preferences on different dimensions of living conditions.

sustainability levels for each country are obtained and those are aggregated to obtain the final FEEM SI values.

This approach gives an innovative direction to the current literature on aggregate indicators. For example in a recent review, Singh et al. (2009) summarises forty-one sustainability indicators and majority of those indices are either aggregated through equal weight assignment (e.g., Environmental Sustainability Index, Human Development Index, Sustainability Performance Index, etc.) or weights given by experts (e.g., Index of Environmental Friendliness) to each sustainability indicator. However, none of those indices allows for the interactions between different sustainability indicators. In other words, those aggregation methodologies do not account for synergies or redundancies when indicators are aggregated. In the construction of FEEM SI, the Choquet integral aggregation is able to address specifically the inter-relations across indicators, thus overcoming the limitations of other aggregation methodologies.

In addition, the questionnaire tailored to elicit experts' evaluations of the sustainability indicators also releases important key characteristics where one can obtain information about the experts' attitude towards the sustainability concept. For example, one of the key aspects that can be derived through the Choquet integral is the "andness" degree. An "andness" degree close to 1 indicates that the decision maker tends to be non-compensative, meaning that she/he would not accept that a good performance in one indicator compensates for a negative one in another. On the contrary, an "andness" degree close to 0 indicates that the decision maker is satisfied even if only one indicator is at "best" level. Given the nature of the problem at hand, it seems more likely that decision makers evaluating the hierarchical structure of the FEEM SI tree should be more inclined towards "andness", as sustainability implicitly requires a balanced development across its different components.[6]

5 Baseline Scenario and Sustainability

The framework described in the previous sections has been applied to a baseline scenario for the period 2005–2020, which gives insights on the evolution of overall sustainability as well as its pillars in a no policy scenario. This scenario is used as reference to analyse the effect of alternative policy scenarios. The baseline scenario replicates the historical trends of main economic variables in the period 2004–2009 and then reproduces an intermediate growth level scenario. The main sources of exogenous dynamic are presented in Table 3. In order to create this scenario the baseline is built according to a set of exogenous drivers, mainly population, labour stock and land productivity. Additional variables such as labour

[6] For a detailed description of the aggregation procedure see Cruciani et al. (2012).

Table 3 Main variables and reference sources in the baseline scenario

Variable	Reference source
Population	UN world population prospect (2010 revision)—medium fertility variant
Fossil fuel prices	Eurelectric (2011)
GDP	2005–2009 = WDI World Bank (2010a) 2010–2020 = MMC_G10 scenario med pop—medium growth—fast convergence (Conv) developed within the RoSE project[a] + IMF (2010) for downscaling at country level
Energy intensity	2005–2009 = IEA (2010) 2010–2020 = endogenous
CO_2 emissions	2005–2009 = IEA (2010) 2010–2020 = endogenous
Public debt	IMF (2010)

[a] "RoSE—Roadmaps towards Sustainable Energy Futures: A Model-Based Assessment of Scenarios for Decarbonising the Energy System in the twenty first Century". Germany

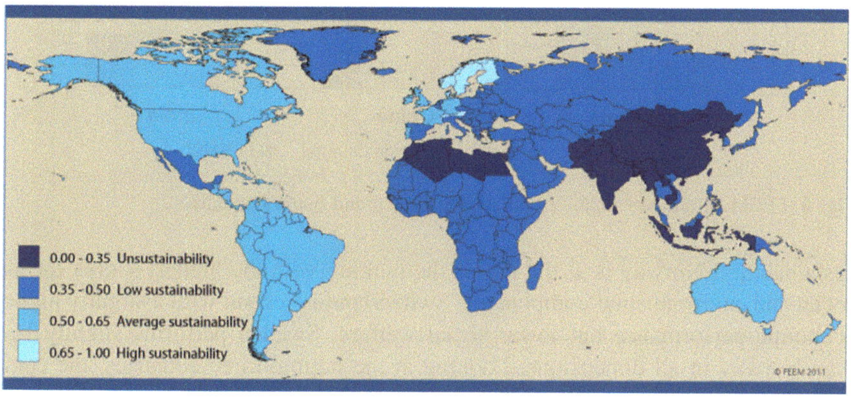

Fig. 3 World map of sustainability in 2011

productivity and total factor productivity are then calibrated to replicate the selected reference GDP growth rate.[7]

The FEEM SI and its indicators are then calculated for each country/macro-region and for each year until 2020. The map (Fig. 3) represents the global picture for the world in 2011. As expected, the most developed countries show higher sustainability than less developed ones. This is mainly explained with the good performances of rich countries in the social pillar.

Figure 4 compares the scores of each pillar (economic, social and environmental) and the aggregate index for the best and worst countries. The scores for the top-three countries are similarly high in the three main components of

[7] The baseline calibration and validation is detailed in Chap. 5 of the FEEM SI Methodological report (FEEM 2011).

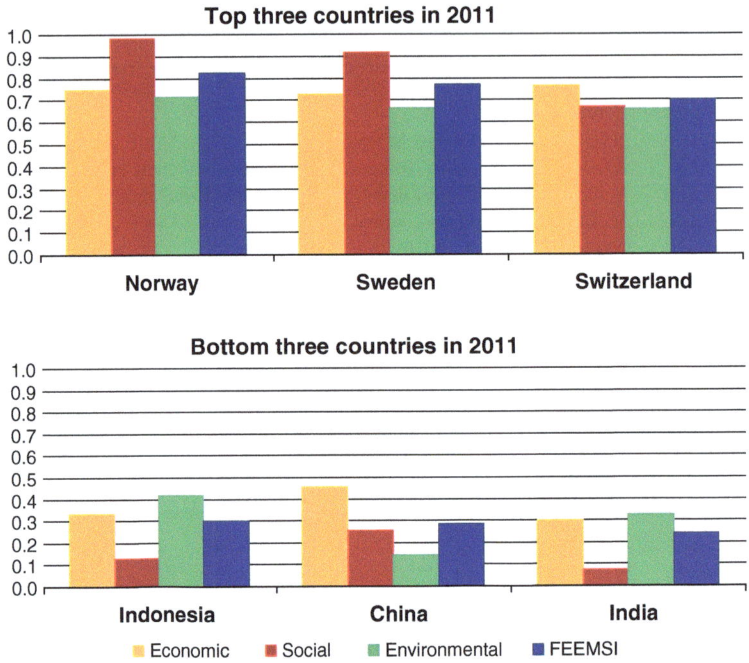

Fig. 4 FEEM SI and sustainability pillars for the top and bottom countries

sustainability. Norway is at the top of the ranking with the highest scores for the social and environmental components. Switzerland is second with a slightly higher economic performance but lower social welfare. Sweden performs slightly less than Norway in all dimensions. Looking at the bottom-three countries, the components are very unequally distributed. Indonesia has a higher value for the environmental dimension than the other two regions. On the other side, China has the highest score in the values of economic and social pillars, while reaching the lowest score in the environmental one. Finally, India reaches the lowest levels in the score of economic and social pillars.

Table 4 illustrates the position of the 40 countries/macro-regions in 2011 and 2020, as well as the changes in the ranking. The results illustrate that no dramatic changes occur in the period under consideration. Benelux (+7 positions from 2011 to 2020), Germany (+5) and Italy (+3) benefit the highest advancements in the sustainability ranking; conversely, United States (−6) and Russia (−5) downgrade mostly, along with a reduction in their overall level of sustainability, since their economic growth determines a significant deterioration of the environmental pillar.

The purpose of a Sustainability Index is to consider economic, social and environmental indicators simultaneously and offer additional and more complete information for welfare assessment beyond what GDP per capita can do. Figure 5 sketches the correlation between GDP p.c. and the FEEM SI. On average, the higher the GDP p.c., the higher the value of FEEM SI. However, the sustainability

Table 4 World sustainability ranking (2020 with respect to 2011)

Rank 2011	Country	FEEM SI 2011	Δ Rank	FEEM SI 2020	Country	Rank 2020
1	Norway	0.82	=	0.85	Norway	1
2	Sweden	0.77	=	0.81	Sweden	2
3	Switzerland	0.70	−1	0.74	Austria	3
4	Austria	0.69	1	0.70	Switzerland	4
5	Finland	0.66	=	0.68	Finland	5
6	Denmark	0.65	=	0.68	Denmark	6
7	Canada	0.64	=	0.67	Canada	7
8	France	0.63	=	0.65	France	8
9	Ireland	0.62	−1	0.63	New Zealand	9
10	New Zealand	0.61	1	0.62	Ireland	10
11	USA	0.55	−6	0.58	Germany	11
12	Australia	0.55	=	0.58	Australia	12
13	Brazil	0.55	−2	0.56	Benelux	13
14	UK	0.53	=	0.55	UK	14
15	RoEurope	0.53	−1	0.54	Brazil	15
16	Germany	0.53	5	0.54	RoEurope	16
17	Portugal	0.52	−2	0.53	USA	17
18	RoLA	0.51	=	0.53	RoLA	18
19	Spain	0.50	−2	0.53	Portugal	19
20	Benelux	0.50	7	0.51	RoEU	20
21	Russia	0.49	−5	0.50	Spain	21
22	RoEU	0.49	2	0.50	Italy	22
23	Mexico	0.49	−2	0.49	Korea	23
24	Korea	0.48	1	0.49	Japan	24
25	Italy	0.47	3	0.48	Mexico	25
26	Japan	0.46	2	0.48	Russia	26
27	Turkey	0.45	=	0.48	Turkey	27
28	MiddleEast	0.45	=	0.47	MiddleEast	28
29	Poland	0.43	=	0.44	Poland	29
30	SouthAfrica	0.43	=	0.43	SouthAfrica	30
31	Greece	0.40	=	0.43	Greece	31
32	RoAfrica	0.40	=	0.40	RoAfrica	32
33	RoWorld	0.39	=	0.39	RoWorld	33
34	SEastAsia	0.37	=	0.36	SEastAsia	34
35	RoFSU	0.37	=	0.36	RoFSU	35
36	NorthAfrica	0.34	=	0.34	NorthAfrica	36
37	RoAsia	0.33	=	0.34	RoAsia	37
38	Indonesia	0.30	−1	0.32	China	38
39	China	0.29	1	0.32	Indonesia	39
40	India	0.24	=	0.29	India	40

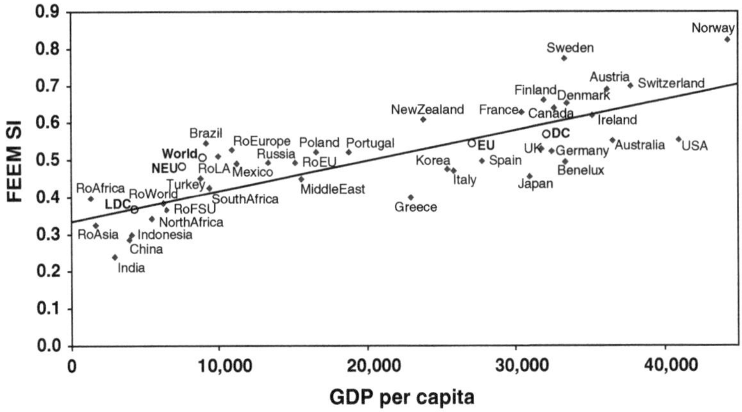

Fig. 5 FEEM SI and GDP p.c. correlation

performance of countries with similar GDP p.c., such as Benelux and Sweden, can be very different.

Differences emerge in comparing the ranking of GDP p.c. and of the FEEM SI. For example, USA and Australia, with the 2nd and 4th highest GDP p.c. in the world respectively, are only at 11th and 12th positions according to the FEEM SI ranking. This is due to the low performance in environmental sustainability not compensated by the good economic and social performance. Other rich countries are significantly worse off when looking at FEEM SI value, such as Japan, Italy and Greece, while Sweden, Finland, France have the reverse relationship (FEEM SI makes them better off than GDP ranking). A stronger relation between GDP p.c. and FEEM SI rankings characterises the 10 bottom countries; a low GDP p.c. is normally associated to a low overall sustainability performance. Nevertheless, the other indicators considered in the FEEM SI skew the GDP p.c. ranking. For instance, India (38th according to GDP p.c.) becomes the worst performer (40th according to FEEM SI) because of its poor performance in social and environmental sustainability. Conversely, the Rest of Africa (RoAfrica) benefits from the relatively good environmental performance connected to the relatively low importance of energy-intensive industry.

6 The Effect of Climate Policy on Sustainability

Climate change is one of the main challenges for humankind in this century. Designing and implementing an effective climate policy offers a valid option to deal with this phenomenon. Nevertheless, curbing CO_2 emissions implies economic costs that often discourage a binding commitment in this field. The FEEM

Table 5 CO_2 emissions growth and reduction targets in 2020 with respect to 1990

Region	Baseline CO_2 growth (%)	CO_2 target (%)	Policy scenario CO_2 growth (%)
Annex I—Leading Regions			
Australia	62	13	13
New Zealand	102	−10	−10
Japan	21	−25	−25
EU27	2	−20	−20
USA	36	−3	−3
Canada	26	3	3
Switzerland	15	−20	−20
Norway	32	−30	−30
Russia	9	−15	−15
Turkey[a]	123	–	191
Non-annex I—leading regions			
Korea (Rep. of)	207	115	115
China[b]	376	–	375
India[b]	367	–	357
Indonesia	335	222	222
Mexico	108	46	46
Brazil	279	142	142
South Africa	83	20	20
Annex I	21	−12	−10
Non-annex I	317	289	289
Rest of the World	115	–	155
WORLD	94	–	75

[a] Annex I country with no target
[b] China and India's targets are originally stated in terms of carbon intensity reduction with respect to 2005

SI, reflecting the broad concept of sustainability, allows analysing the benefits of a climate policy in a more comprehensive way.

The analysis focuses on a mitigation scenario in which Annex I and Non-Annex I countries taking action towards climate change achieve the low pledges proposed at the 15th UNFCCC Conference of the Parties in Copenhagen (December 2009). All countries implement a unilateral emission reduction through a carbon tax or a carbon intensity target (China and India). The only exception is represented by EU27, whose Member States are allowed trading emission permits among them (but not with the rest of the world) replicating the Emission Trading Scheme in force since 2005. For sake of simplicity, the policy only refers to CO_2 emissions and is applied uniformly to all productive sectors. Table 5 reports the Copenhagen targets, and percentage change in both baseline and policy case for leading countries.

Looking at the main aggregates in Table 5, Annex I countries, which in the baseline scenario increase emissions in 2020 by 21 % with respect to 1990, reduce their emission levels by 10 % in the policy scenario. Non Annex I countries also contribute to the policy since their emissions grow less than in the baseline (289 vs. 317 %). The Rest of the World, with no commitments, increases its emissions

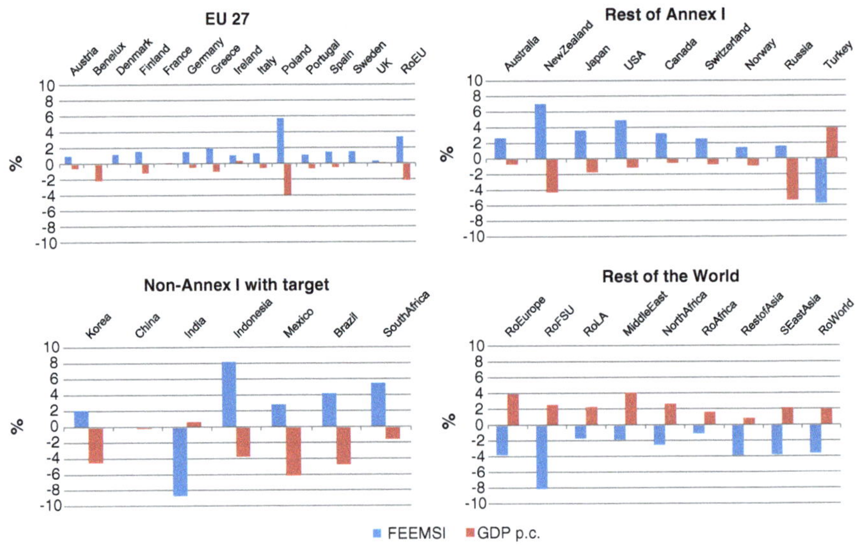

Fig. 6 FEEM SI and GDP p.c. % change in 2020 (climate policy vs. baseline)

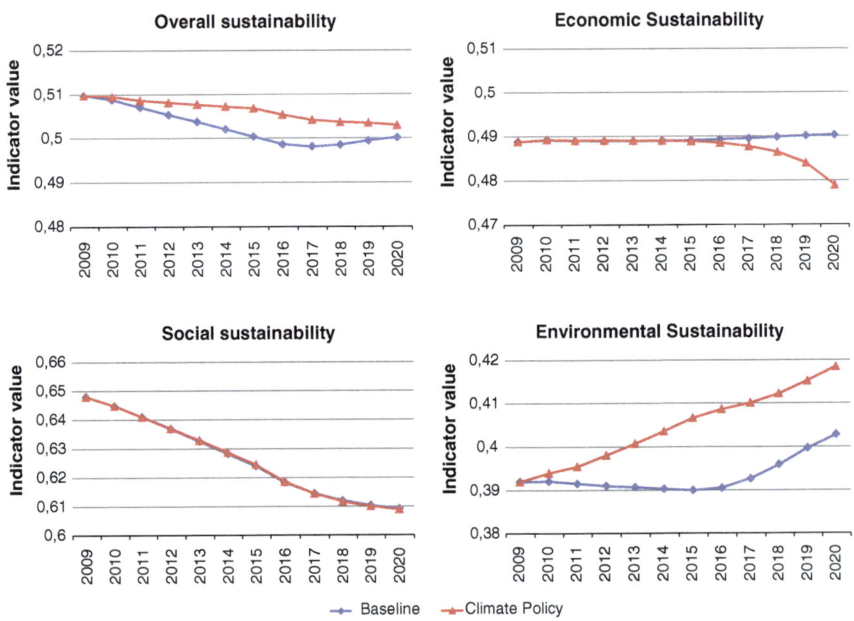

Fig. 7 Changes in World sustainability by pillar (climate policy vs. baseline)

from 115 to 155 %. At world level, emissions after the mitigation policy are lower than in the baseline scenario, growing 75 % instead of 94 %.

Figure 6 shows the implications of the climate policy for sustainability and mitigation costs of several aggregates. In EU27, Poland and RoEU display the main GDP losses, but also the highest improvement in sustainability. These two countries contribute more than the others to the EU abatement, given their low mitigation costs. Benelux also has a significant economic loss, but in this case the impact on sustainability is negligible. Germany, Sweden and Ireland show an increase in sustainability at very low cost, given the already good environmental performance. Among other Annex I countries, the highest costs are undertaken by Russia and New Zealand.

The related positive impact on sustainability is differentiated: high for New Zealand but quite low for Russia. USA, Australia and Canada have a significant increase in sustainability with low economic loss, meaning once again that the initial stage of technological development matters. Turkey not having any commitment would experience an improvement of economic conditions, but with a substantial reduction in its sustainability due to the increased environmental degradation.

Almost all Non-Annex I countries show important economic costs to achieve their own targets (especially Mexico, Brazil and Korea). Indonesia has the strongest increase in sustainability. India earns in GDP terms but with a drop in sustainability, while China has a negligible loss with no impact on sustainability. In both cases the economic result depends on lack of stringency of the target (almost achieved in the baseline). Overall, costs are higher for Non-Annex I than for Annex I countries. Rest of the World macro-regions are all better off with respect of GDP since they do not have any emissions target and can increase their output due to the carbon leakage effect; but at the same time their sustainability decreases due to environment degradation.

The implication of the climate policy for sustainability at world level by pillar is depicted in Fig. 7. The overall sustainability declines less than in the baseline scenario. The downward trend is justified by the significant decrease in the social pillar (as in the baseline), almost unaffected by climate policy. However, the increase in the environmental pillar more than compensates the decline in the economic pillar after 2015, when the policy becomes more costly. The mitigation policy improves world sustainability. Moreover, this positive result could be stronger if a higher number of signatories committed to an emission reduction target, reducing the carbon leakage effect.

7 Conclusions

This chapter presented a methodological tool for sustainability measurement built in a CGE model: the FEEM SI. Most policy-makers and stakeholders recognise the importance to go beyond the economic dimension in measuring sustainable development. While many highlight the opportunity to change the development pattern through qualitative approaches, there is an increasing interest in quantifying the level of sustainable development.

The FEEM SI summarises a set of indicators reflecting the main aspects of sustainability. It uses a normalisation procedure based on re-scaling and benchmarking to reconcile all indicators to a common scale. The indicators' aggregation requires the elicitation of experts' evaluations through an "ad hoc" questionnaire in order to derive weights, and a non-linear aggregation procedure of weights and indicators values.

The FEEM SI offers projections on the trend of countries' sustainability across the world in the next future and allows considering different scenarios besides the current situation. This requires the use of a recursive-dynamic CGE model as basic framework for the index in which the overall coherence is guaranteed by economic interrelations among countries.

The FEEM SI results show a heterogeneous situation, in which advanced economies have a satisfying level of sustainability while developing countries still show a significant gap. Looking in detail at the determinants of this result, it emerges that a high performance in each sustainability dimension is a necessary condition to reach the overall sustainability.

In the baseline scenario, world sustainability slightly decreases mainly due to a significant reduction in the economic and social components. In the climate policy scenario, sustainability in signatory countries increases since the costs and the subsequent reduction in economic performance are more than offset by the improvement of the state of environment. Both mitigation effectiveness and sustainability at world level can be seriously compromised because of carbon leakage. These results suggest that a higher level of sustainability could be achieved if a higher number of signatories committed to an emission reduction target.

Acknowledgments This chapter is part of the research of the Climate Change and Sustainable Development Research Programme of the Fondazione Eni Enrico Mattei. The FEEM Sustainability Index has benefited from support from researchers outside the FEEM SI team as well as the contribution of a set of experts who responded a questionnaire. We would like to thank them for their patience and help. The authors would also like to acknowledge the anonymous referee for the useful comments provided.

A.1 Annex I

See Tables 6, 7 and 8

Table 6 Indicators' description

Dimension	Name	Indicator	Long description	Literature reference
Economic	R&D	R&D expenditure/GDP (%)	This indicator assumes a positive relationship between investment in R&D and growth, by maintaining that increased investment in R&D can bring more R&D output that will eventually lead to more innovation and increased productivity	EU SDS—UN CSD—WDI
	Investment	Net investment/capital stock (%)	Investment is one of the main drivers of economic sustainability, allowing for capital accumulation, which boosts economic growth. This indicator is weighted considering the country specific capital stock	EU SDS—UN CSD
	GDP per capita	GDP PPP/population	It is a measure of the per capita value of all market goods and services produced within a country. GDP p.c. is the typical indicator used to define the average well-being in a country	EU SDS—UN CSD—WDI
	Relative trade balance	Trade balance/market openness	The relative trade balance measures the degree of a country's exposure in the global commodities markets. It considers the net export value and weights it with the country specific market openness (exports + imports). Relying relatively more upon exports is a signal of strong competitiveness	–
	Public debt	Government debt/GDP (%)	Public debt has an important role on the future perspective of a country's economy. It depends on current government choices on expenditure and taxation, and on previously accumulated debt	WDI—UN CSD—IMF
Social	Population density	Population/country surface	Population density evaluates the population concentration in a specific country or macro-region (excluding uninhabitable areas). It represents the pressure on the available living space and resources for each individual	UN CSD—WDI
	Education	Education exp./GDP (%)	Expenditure in education constitutes an investment in human capital. The role of education in improving future economic conditions and enhancing mobility as well as gender equality is supported by several studies	EU SDS—WDI
	Health	Total health exp./GDP (%)	The generalised access to basic health services is a major concern throughout the world. Monitoring the growth of expenditures in health by summing public and private expenditures allows to measure the degree of support on this issue	WDI
	Food relevance	Food cons./private exp. (%)	This indicator is used as a proxy for the poverty level. In fact, according to Engel's law, the higher the proportion of national income spent on food the lower the level of a country's welfare	–
	Energy imported	Energy imported/energy cons. (%)	This is an indicator of energy security. The higher the energy dependence from abroad, the higher the risks deriving from changes in energy prices and political instability in energy-rich countries	WDI
	Energy access	Population with access to electricity/total population (%)	Access to energy is important with reference to living conditions and future prospectives of well-being. This indicator considers the share of population having access to electricity. It allows capturing the intra country aspect of energy security, being more focused on distribution of energy resources than on availability at the country level	WDI
	Private health	Private health exp./total health exp.(%)	Monitoring the balance between public and private contribution to the health sector is essential for sustainability because it determines the availability of primary service to the whole society. The higher the share of private health expenditure, the lower the ability of poorer people to access to the health care	WDI

(continued)

Table 6 (continued)

Dimension	Name	Indicator	Long description	Literature reference
Environmental	GHG per capita	Kyoto GHGs emissions/population	The greenhouse gases are considered as described in the Annex I of the Kyoto Protocol. Emission per capita is a measure of the burden that the society imposes on climate and environment	EU SDS—UN CSD—WDI
	CO_2 intensity	CO_2 emissions/total primary energy cons.	This indicator is fundamental to monitor the improvement of the environmental performance of production and consumption activities, the latter playing a major role in the release of carbon dioxide into the atmosphere	EU SDS—UN CSD—WDI
	Energy intensity	Total primary energy supply/GDP PPP	This indicator aims to assess the evolution of energy use efficiency	EU SDS—UN CSD—WDI
	Renewables	Renewable cons./total primary energy cons.(%)	The gradual reduction of fossil fuel use is an important step towards security and sustainability of energy systems. The higher the share of green energy, the higher the environmental performance of the energy sectors	EU SDS—UN CSD—WDI
	Plants	Endangered species/total species (%)	This indicator represents an alarm signal of the general worsening of habitats. It provides a comparable measure of endangered Plant species throughout the world, by considering the number of endangered species over the number of total known species present in that country	EU SDS—UN CSD—WDI
	Animals	Endangered species/total species (%)	As in the previous indicator, it also represents an alarm signal of the general worsening of habitats. It is calculated in the same way but focusing on animal biodiversity	EU SDS—UN CSD—WDI
	Water	Water use/total available water (%)	Human pressure on water, is an important indicator of resource pressure. It is estimated as water consumed in a country (for agriculture, industry and private uses) over the total renewable water resources available in that specific country	UN CSD—WDI

Table 7 Regional aggregation

No.	Macro-Regions	Countries
1	Australia	Australia
2	NewZealand	New Zealand
3	Japan	Japan
4	Korea	Korea
5	China	China, Hong Kong, Taiwan
6	India	India
7	Indonesia	Indonesia
8	SEastAsia	Malaysia, Philippines, Singapore, Thailand, Vietnam
9	RoAsia	Afghanistan, Bangladesh, Bhutan, Brunei Darassalam, Cambodia, Democratic Republic of Korea, Lao People's Democratic Republic, Macau, Maldives, Mongolia, Myanmar, Nepal, Pakistan, Sri Lanka, Timor East
10	USA	USA
11	Canada	Canada
12	Mexico	Mexico
13	Brazil	Brazil
14	RoLA	Argentina, Bolivia, Chile, Colombia, Ecuador, Paraguay, Peru, Uruguay, Venezuela, Falkland Islands (Malvinas), French Guiana, Guyana, Suriname, Costa Rica, Guatemala, Nicaragua, Panama, Belize, El Salvador, Honduras, Saint Vincent and the Grenadines, Trinidad and Tobago, Turks and Caicos, Anguilla, Antigua & Barbuda, Aruba, Bahamas, Barbados, Cayman Islands, Cuba, Dominica, Dominican Republic, Grenada, Guadeloupe, Haiti, Jamaica, Martinique, Montserrat, Netherlands Antilles, Puerto Rico, Saint Kitts and Nevis, Saint Lucia, Virgin Islands (British), Virgin Islands (U.S.)
15	Austria	Austria
16	Benelux	Belgium, Luxembourg, Netherlands
17	Denmark	Denmark
18	Finland	Finland
19	France	France
20	Germany	Germany
21	Greece	Greece
22	Ireland	Ireland
23	Italy	Italy
24	Poland	Poland
25	Portugal	Portugal
26	Spain	Spain
27	Sweden	Sweden
28	UK	UK
29	RoEU	Cyprus, Czech Republic, Estonia, Hungary, Latvia, Lithuania, Malta, Slovakia, Slovenia, Bulgaria, Romania
30	Switzerland	Switzerland
31	Norway	Norway
32	RoEurope	Albania, Andorra, Bosnia and Herzegovina, Croatia, Faroe Islands, Gibraltar, Iceland, Liechtenstein, Macedonia, the former Yugoslav Republic of, Monaco, San Marino, Serbia and Montenegro

(continued)

Table 7 (continued)

No.	Macro-Regions	Countries
33	Russia	Russia
34	RoFSU	Belarus, Ukraine, Moldova, Republic of, Kazakhstan, Kyrgyzstan, Tajikistan, Turkmenistan, Uzbekistan, Armenia, Azerbaijan, Georgia
35	Turkey	Turkey
36	MiddleEast	Bahrain, Islamic Republic of Iran, Iraq, Israel, Jordan, Kuwait, Lebanon, Occupied Palestinian Territory, Oman, Qatar, Saudi Arabia, Syrian Arab Republic, United Arab Emirates, Yemen
37	NorthAfrica	Algeria, Egypt, Libyan Arab Jamahiriya, Morocco, Tunisia
38	RoAfrica	Angola, Benin, Botswana, Burkina Faso, Burundi, Cameroon, Cape Verde, Central African Republic, Chad, Comoros, Congo, Democratic Republic of the Congo, Cote d'Ivoire, Djibouti, Equatorial Guinea, Eritrea, Ethiopia, Gabon, Gambia, Ghana, Guinea, Guinea-Bissau, Kenya, Lesotho, Liberia, Madagascar, Malawi, Mali, Mauritania, Mauritius, Mayotte, Mozambique, Niger, Nigeria, Reunion, Rwanda, Saint Helena, Sao Tome and Principe, Senegal, Seychelles, Sierra Leone, Somalia, Sudan, Swaziland, Tanzania, Togo, Uganda, Zambia, Zimbabwe
39	SouthAfrica	SouthAfrica
40	RoWorld	American Samoa, Cook Islands, Fiji, French Polynesia, Guam, Kiribati, Marshall Islands, Micronesia, Federated States of, Nauru, New Caledonia, Norfolk Island, Northern Mariana Islands, Niue, Palau, Papua New Guinea, Samoa, Solomon Islands, Tokelau, Tonga, Tuvalu, Vanuatu, Island of Wallis and Futuna, Bermuda, Greenland, Saint Pierre and Miquelon

Table 8 Sectoral aggregation

No	Sectors
1	Food
2	Forestry
3	Fishing
4	Coal
5	Oil
6	Gas
7	Petroleum products
8	Other electricity
9	Renewables
10	Nuclear
11	Biofuels
12	Energy intensive industries
13	Other industries
14	Water
15	Market services
16	Public services
17	R&D
18	Education
19	Private health
20	Public health

References

Böhringer C, Löschel A (2006) Computable general equilibrium models for sustainability impact assessment: status quo and prospects. Ecol Econ 60(2006):49–64

Carraro C, Campagnolo L, Eboli F, Lanzi E, Parrado R, Portale E (2012) Quantifying sustainability: a new approach and world ranking. In: Working Papers 2012.94. Fondazione Eni Enrico Mattei, Milan

Commission on the Measurement of Economic Performance and Social Progress (2009) Report of the commission on the economic and social progress. http://www.stiglitz-sen-fitoussi.fr/documents/rapport_anglais.pdf. Accessed 15 Nov 2012

Cruciani C, Giove S, Pinar M, Sostero M (2012) Constructing the FEEM sustainability index: a choquet-integral application. In: Working papers 2012.50. Fondazione Eni Enrico Mattei, Milan

Daly HE, Cobb JJ (1989) For the common good: redirecting the economy toward community, the environment and a sustainable future. Beacon Press, Boston

Eurelectric (2011) Power choices—pathways to carbon-neutral electricity in Europe by 2050. http://www.eurelectric.org/PublicDoc.asp?ID=63875. Accessed 15 Nov 2012

European Commission (2008) Energy sources, production costs and performance of technologies for power generation, heating and transport. EU Energy Secur Solidarity Action Plan. SEC (2008) 2872

FAO and IIASA (2000) Global agro-ecological zones. http://www.iiasa.ac.at/Research/LUC/GAEZ/index.htm. Accessed 10 Feb 2011

FEEM (2011) FEEM Sustainability Index Methodological Report 2011. Fondazione Eni Enrico Mattei: http://www.feemsi.org/pag/downloads_2.php

Food and Agriculture Organization of the United Nations (FAO) (2011) Aquastat database. http://www.fao.org/nr/water/aquastat/data/query/index.html?lang=en. Accessed 10 Feb 2011

Grabisch M (1995) Fuzzy integral in multicriteria decision making. Fuzzy Sets Syst 69:279–298

Grabisch M (1996) The application of fuzzy integrals in multicriteria decision making. Eur J Oper Res 89:445–456

Grabisch M, Labreuche C, Vansnick JC (2003) On the extension of pseudo-Boolean functions for the aggregation of interacting criteria. Eur J Oper Res 148(1):28–47

GRI—Global Reporting Initiative (2006) Sustainability reporting guidelines, version 3.1. Global Reporting Initiative, Amsterdam

GRI—Global Reporting Initiative (2010) Global reporting initiative sustainability report 2009/2010, Amsterdam

GTZ (2009) Energy-policy framework conditions for electricity markets and renewable energies. Eschborn

Hertel TW (ed) (1997) Global trade analysis: modeling and applications. Cambridge University Press, Cambridge

Horridge M (2008) SplitCom: programs to disaggregate a GTAP sector. Centre of Policy Studies, Monash University, Melbourne

IEA (2005) Projected costs of generating electricity. International Energy Agency, Paris

IEA (2010) World Energy Outlook 2010. International Energy Agency, Paris

IMF (2010) World economic outlook 2010 rebalancing growth. International Monetary Fund, Washington

IUCN—International Union for Conservation of Nature and Natural Resources (2010) Red list of threatened species. http://www.iucnredlist.org/. Accessed 10 Feb 2011

Krajnc D, Glavic P (2005) A model for integrated assessment of sustainable development. Resour Conserv Recycl 43:189–208

Lee H (2008) The combustion based CO2 emission data for GTAP version 7 data base. Center for Global Trade Analysis, Purdue University, West Lafayette

Marichal J-L, Roubens M (2000) Determination of weights of interacting criteria from a reference set. Eur J Oper Res 124(3):641–650

Meyer P, Ponthière G (2011) Eliciting preferences on multiattribute societies with a Choquet integral. Comput Econ 37(2):133–168
Murofushi T, Soneda S (1993) Techniques for reading fuzzy measures (III): interaction index. In: 9th fuzzy system symposium. Sapporo, pp 693–696
Murofushi T, Sugeno M, Machida M (1994) Non-monotonic fuzzy measures and Choquet integral. Fuzzy Sets Syst 64:73–86
Narayanan B, Walmsley TL (eds) (2008) Global trade, assistance, and production: the GTAP 7 data base. Center for Global Trade Analysis, Purdue University, West Lafayette
OECD (2008) Handbook on constructing composite indicators: methodology and user guide. OECD Publishing, Paris
Parris TM, Kates RW (2003) Characterizing and measuring sustainable development. Ann Rev Environ Resour 28(1):559–586
Ragwitz M, Resch G, Morthorst PM, Coenraads R, Konstantinaviciute I, Heyder B (2007) OPTRES: assessment and optimisation of renewable support schemes in the European electricity market, Intelligent Energy Europe. http://www.optres.fhg.de/OPTRES_FINAL_REPORT.pdf
REN21 (2011) Renewable energy policy network for the 21st century. http://www.ren21.net. Accessed 10 Feb 2011
Rose SK, Lee H (2008) Non-CO_2 greenhouse gas emissions data for climate change economic analysis. In: GTAP working paper no. 43
Singh RK, Murty H, Gupta S, Dikshit A (2009) An overview of sustainability assessment methodologies. Ecol Ind 9(2):189–212
Thomas CD, Cameron A, Green RE, Bakkenes M, Beaumont LJ, Collingham YC, Erasmus BFN, de Ferreira Siqueira M, Grainger A, Hannah L, Hughes L, Huntley B, van Jaarsveld AS, Midgley GF, Miles L, Ortega-Huerta MA, Townsend Peterson A, Phillips OL, Williams SE (2004) Extinction risk from climate change. Nature 427:145–148
UN CSD (2005) Indicators of sustainable development—CSD theme indicator framework, New York
UN (2010) World population prospect. http://www.un.org/esa/population/. Accessed 10 July 2011
UN (2012) The future we want. Draft resolution submitted by the President of the general assembly, Sixty-sixth session, Agenda item 19, Sustainable development A/66/L.56, 24 July 2012
UNDP (1990) Human development report. http://www.undp.org. Accessed 10 July 2011
WCED—World Commission on Environment and Development (1987) Our common future. Published as annex to general assembly document A/42/427, Development and international co-operation: environment August 2, 1987
World Bank (2010a) World development indicators. http://data.worldbank.org/indicator. Accessed 10 July 2011
World Bank (2010b) Addressing the electricity access gap. Background paper for the World Bank group energy sector strategy, World Bank
World Health Organization (2010) http://apps.who.int/ghodata/?vid=1901. Accessed 10 July 2011
Yusuf JA, El Serafy S, Lutz E (1989) Environmental accounting for sustainable development. In: UNEP World Bank symposium, Washington
Yale and Columbia Universities (2010) 2010 environmental performance index, summary for policymakers. http://epi.yale.edu/. Accessed 10 July 2011

Measuring and Evaluating Business Sustainability: Development and Application of Corporate Index of Sustainability Performance

Frank Medel-González, Lourdes García-Ávila, Adael Acosta-Beltrán and Cecilia Hernández

Abstract In the last 20 years, from the Rio Summit, has been a growing concern for global sustainability in all sectors of society. The business organizations do not escape this trend and seek more sustainable ways to generate value. This phenomenon has been driven primarily by the associated legislation arising from the need to conserve natural resources and reduce impacts across economic, social and environmental dimensions, associated with organizations performance. This research proposes a structured approach for sustainability performance evaluation trough *Corporate Index of Sustainability Performance* (*CISP*) in Cuban organizations, combining different tools like: ISO 14031, Sustainability Reporting Guidelines of Global Reporting Initiative, Balanced Scorecard and muticriteria methods. Also is exposed the design of a web application that enable the management, storage and integration of sustainability indicators and CISP calculus for assessing the business sustainability performance. Were used, as study case; four small power plants of distributed generation in electric sector of Villa Clara, Cuba.

Keywords Corporate sustainability · Performance measurement · Multicriteria methods

F. Medel-González (✉) · L. García-Ávila
Industrial Engineering Department, Universidad Central "Marta Abreu" de Las Villas, Santa Clara, Cuba
e-mail: frankmedel@uclv.edu.cu

L. García-Ávila
e-mail: Lourdes@uclv.edu.cu

A. Acosta-Beltrán
Information Technologies Department, GEOCUBA VCL-SSP, Santa Clara, Cuba
e-mail: abeltran@vcl.geocuba.cu

C. Hernández
Production Engineering Department, Fluminense Federal University, Rio the Janeiro, Brazil
e-mail: ctoledo@id.uff.br

M. G. Erechtchoukova et al. (eds.), *Sustainability Appraisal: Quantitative Methods and Mathematical Techniques for Environmental Performance Evaluation*, EcoProduction, DOI: 10.1007/978-3-642-32081-1_3,

1 Introduction

Since 1987, the definition of Sustainable Development has been appreciated in the international arena, in all sectors of the economy, awareness related sustainability. This phenomenon has been driven primarily by the associated legislation arising from the need to conserve natural resources and reduce impacts across economic, social and environmental dimensions, associated with organizations performance.

In recent years have been arising different business reporting models that guide companies to understand, demonstrate, communicate, report and improve their sustainability performance, such as, Eco-Management and Audit Schema, International Standard Organization (ISO 14000 series) and Global Reporting Initiative (GRI).

However, in Cuba, gaps remains in relation to corporate sustainability performance measurements and evaluation; as internal management process that helps organizations select, collect, integrate and evaluate sustainability indicators. These indicators should respond to the policies, strategies and goals of the organizations according to their business area; providing key information for the corporate sustainability decision making process.

The main goal of this research is propose a structured approach to make operative the sustainability performance measurement and evaluation process in Cuban organizations, combining different tools like: ISO 14031, Sustainability Reporting Guidelines of Global Reporting Initiative, Balanced Scorecard one of the most popular managerial tools, that link performance measurement to strategy, using a multidimensional set of financial and non-financial performance metrics (Bonacchi and Rinaldi 2007) and Analytic Hierarchy Process (AHP) and Analytic Network Process (ANP), the most used multi-criteria decision-making methods in the last 20 years.

The chapter presents an effective contribution in sustainability performance measurement and evaluation process, making it operational through a *Corporate Index of Sustainability Performance* (*CISP*). The *CISP* is a numerical and descriptive categorization of a large amount of information, in order to simplify evidence contained in triple bottom line indicators.

The literature showed a strong tendency to composite index construction in environmental and sustainability areas (Puolamaa et al. 1996; Cherchye and Khuosmanen 2002; Chiang and Lai 2002; Damjan and Glavic 2005; Castellanos-Abella and Van Westen 2007; Gómez-Limón and Riesgo 2008; Blanc et al. 2008; Sellito et al. 2010; Broche-Fernández and Ramos-Gómez 2010) but mechanisms are lacking in order to measure and evaluate corporate sustainability performance in objectives terms, focusing on a single numerical index. It could offer decision makers condensed information for progress evaluation and benchmarking comparisons, and make decision making more quantitative, empirically grounded, and systematic (Esty et al. 2005) and helps organizations to illustrate progress and setbacks in relation to organizational sustainability performance and identify the critical areas.

Information Technology can play an important role in sustainability management, specifically in the evaluation of sustainability performance. A practical contribution is exposed, a web application that enable the management, storage and integration of sustainable indicators; the basis for assessing the corporate sustainability of organizations and support the *CISP* outcome.

The web application allows the generation of reports from the stored information and provides *CISP* analysis, which aims to determine the level of compliance with the efforts of management regarding sustainability goals defined. The web application makes use of various computer technologies such as *MySQL* as database manager, *Zend Framework* of *PHP*, *Propel* Object Relational Mapping (ORM) and *Business Intelligence and Reporting Tools* (*BIRT*) for report generation. The chapter presents the study case results in four electric power plants in Cuba.

2 Corporate Sustainability

The concept of *corporate sustainability* (CS) has therefore grown in recognition and importance because the organizations are trying to balance their performance among economic, environmental and social domains. The traditional organizational performance measurement related shareholder point of view, has change dramatically in the last 20 years; according Hubbard (2009) a more stakeholder-based view has gradually come to prevail; bringing a multidimensional performance measurement system, distributed over different fields and stakeholders interest.

Many definitions have been developed in the literature in relation with corporate sustainability. This effort responds to companies necessities to bring *Sustainable Development* concept into strategies and daily business activities.

CS refers the incorporation of the triple bottom line objectives into company's operational practices; is a multidimensional concept which includes: business strategies, financial returns, costumer's satisfaction, stakeholder's interests, internal process and human factor. Sustainability goals are often broad and to assess performance, organizations must focus on specific issues or areas of priority (Epstein and Marie-Josée 2001). Other concept outlines how the leaders achieve their business goals by gearing their strategies and management to harness the market's potential for sustainability products and services while at the same time successfully reducing and avoiding sustainability costs and risks (Knoepfel 2001).

According Schaltegger and Burritt (2005) CS is a broad approach that includes various characteristics, in particular relating to the contextual integration of economic, environmental and social aspects".

Esterhuyse (2008) define CS as multi-objective concept which includes the following aspects:

Strategy: integrating long-term economic, environmental and social aspects in their business strategies while maintaining global competitiveness and brand reputation.

Financial: meeting shareholder demands for sound financial returns, long-term economic growth, open communications and transparent financial results.

Customer and Products: fostering loyalty by investing in customer and supplier relationship management products and service innovation that focuses on technologies and practices which use financial, natural and social resources in an efficient, effective and economic manner over the long term.

Governance and stakeholder: setting the highest standards of Corporate Governance and stakeholder engagement, including corporate codes of conduct and public reporting.

Human factor: managing human resources to maintain workforce capabilities and employee satisfaction through best-in-class organizational learning, knowledge management, practices, remuneration and benefit progress.

The correct interrelationship and correspondence among these elements and an appropriated sustainability performance system should enable organizations to generate a long-term economic growth based in costumers' satisfaction with products and services, reinforcing stakeholder's engagement with a motivated human capital assuring long-term sustainability business success.

Corporate sustainability requires that management improves corporate economic performance through voluntary, proactive environmental and social activities (Schaltegger and Burritt 2005). According to Kates et al. (2001), the purpose of sustainability assessment is provide decision-makers an evaluation of global to local integrated nature–society systems in short- and long-term perspectives in order to assist them to determine which actions should or should not be taken in an attempt to make society sustainable.

3 Corporate Sustainability Indicators

The business have a big responsibility in the transition process to sustainable development. The business managers should find ways and tools for balance the organizations performance in different dimensions. Tracking their performance in triple-bottom-line, permits evaluate the pertinence of corporate sustainability goals defined and identify gaps and critical points. The legal requirements identification plays an important role in setting goals process, regulatory compliance could serve as first's help for sustainability goal definitions. Other important issue in this challenge is: bring on board the stakeholders interest.

In the last 15 years, more than a hundred standards and management solutions were developed to evaluate and report the economic, social, environmental and sustainability performance of companies like ISO, Advisory Group on Corporate Social Responsibility (Perrini and Tencati 2006). The diversity of existing frameworks could appear as business strength to achieve more sustainable business. Despite this phenomenon has introduce confusion in organizations in relation of ¿how to measure progress in corporate sustainability?, ¿which tools should be used?, ¿which indicators or metrics are better? When these questions are analyzed, one element could be considered common in the three cases: "indicators".

The importance of indicators for measuring business performance has been widely used by managers. Metrics often establish the implementation framework of organizational strategy and enhance the understanding that value could be created.

Sustainable development indicators and composite indicators are considered to be a good vehicle in helping to measure sustainable development and progress achieved in it (UNCSD 2012).

A composite indicator is the compilation of individual indicators into a single index, on the basis of an underlying model of the multidimensional concept that is been measured.

A metric or indicator, to be effective, must be a verifiable measure and must be based on a well understood and documented process. Moreover, it must have reference points, developed internally or externally, that can act as absolute standards (Purba et al. 2006).

In organizations, indicators can be used:

(1) to evaluate and control the performance of resources
(2) to communicate performance to external as well as internal stakeholders
(3) to suggest improvement by identifying gaps that require intervention and improvement.

The indicators facilitate the measurement of sustainability performance and enable the evaluation of main impacts. They provide information for the compilation of the data that needs to be collected based on the regulations and legislation. Thus, the sustainability indicators provide information for communicating with the stakeholders and the authorities (Wessman and Pihkola 2009).

Despite the indices developed, there is still no useful method for integrated sustainability assessment on the company level available. Although the common principle to aggregate indicators for assessment of the company has gained acceptance, it has also become evident that methods for the aggregation of indicators are either not sufficiently well established yet, or are under development, or are not available with respect to all the sustainability aspects (Statistics Finland 2003).

For that reason many organizations are trying to develop new and exhaustive sustainability measurement systems to tracking their business sustainability goals. Currently there is no single, universally accepted definition or assessment metrics for sustainable development. There are no internationally agreed sustainable development indicators that would help monitor progress (UNCSD 2012).

4 Tools for Environmental and Sustainability Performance Measurement

In the last 20 years, had been developed different reporting models around the world, related environmental and sustainability performance. These reporting models had the finality to help tracking environmental and sustainability strategies at all levels.

4.1 Business Reporting Models

4.1.1 ISO 14031

The ISO 14031 refers to the organizational environmental performance evaluation (EPE) as a process and internal management tool, designed to provide direction continuously reliable and verifiable information to determine if the environmental performance of an organization is complying with the criteria established by the managers. This International Standard supports the requirements of ISO 14001 and the guidance given in ISO 14004, but can also be used independently. ISO 14031 provides guidance on the design and use of environmental performance evaluation within an organization. It describes two broad categories of EPE indicators:

1. *Environmental Performance Indicators* (EPI): specific expressions that provide information about the environmental performance of an organization.
2. *Environmental Condition Indicators* (ECI): provide information on the environmental condition. This information can help an organization to understand the actual or potential impact of its environmental aspects, and thus support the planning and implementation of the EPE.

ISO 14031permit the inclusion of stakeholders interest in business management and economic performance associated environmental protection.

4.1.2 Global Reporting Initiative

The GRI is a reporting framework that intended to serve as a generally accepted framework for reporting on an organization's economic, environmental, and social performance. It is designed for use by organizations of any size, sector, or location. This pattern is for voluntary use by organizations desiring to report on the triple bottom line impacts of their activities, products and services. The GRI sets out principles and specific content to help guide the development of sustainability reporting at the organizational level. In this way, it helps the institutions to present a "balance" and reasonable picture of their economic, environmental and social comparison promotes memory and facilitates interaction and communication with a big range of stakeholders.

GRI include the following elements in a report that complements and only draws selectively from the financial statements:

- Vision and strategy.
- Profile.
- Governance structure and management systems.
- Performance indicators.

GRI measures the elements of business sustainability that have not been addressed before, such as product reparability, activities in developing countries and community technology transfer, among others. In addition, GRI addresses key issues of global concern, such as greenhouse gas emissions, persistent organic pollutants, and the gap between developed and developing countries.

4.2 Balanced Scorecard

The Balanced Scorecard (BSC) is one of the most influential management ideas of the past 20 years. This measurement system was proposed, the first time in 1992 in the article The Balance Scorecard—Measures that Drive Performance written by Robert S. Kaplan and David P. Norton and published in the Harvard Business Review.

The evaluation of an organization must not be restricted to traditional financial evaluation rather it should be complemented with measures related to the satisfaction of costumers, internal processes and the capability to innovate. These additional measures should guarantee the financial companies future and lead the company toward their strategic goals while it maintains these four perspectives equilibrated and balanced (Kaplan and Norton 2000).

Several authors have approached how the traditional balanced scorecard can contribute to the sustainable development, defining the Sustainability Balanced Scorecard (*SBSC*) it is develop for the "Business Case", where the environmental and social topics are used to generate economic value, without committing future generations.

A *SBSC* is a type of BSC specifically designed to reflect the issues and objectives of corporate sustainability. In order to clarify appropriate sustainability strategies and translate them into action, it is generally recommended that managers first design a separate *SBSC*. This must then be integrated into the traditional BSC in order to ensure a holistic view of sustainability. This process will help to overcome the distinction between a traditional financially oriented management approach and emphasizing sustainability or environmental management concerns (Figge et al. 2002).

According to Gminder (2005) the *SBSC* is based on the traditional BSC, but provides a broader scope, integrating the three dimensions of sustainability. So, it has a different content and possibly a different structure ("architecture"). In addition to the four perspectives of the traditional BSC, it is possible to include a fifth perspective in order to explicitly address stakeholder issues. Another definition was given for Bieker (2003), which outlines that the *SBSC* can help to detect important environmental and social strategic objectives in the company, in a strategic business unit or department, illustrating the causal relationships, among the intangible factors and the finances of the company.

The inclusion into *SBSC* of stakeholders' interest is very important according to Kaplan and Norton (1996) "All stakeholder interests, when they are vital for the success of the business unit's strategy, can be incorporated in a Balanced Scorecard", for that reason a different architecture is shown in the Fig. 1.

The stakeholders' perspective permits: (1) list the main interest parts of the business who can affect the value chain, (2) the inclusion into the core management of the business of key topics and concerns that have been raised, and (3) how the organization responds to those key topics. The social and cultural perspective allows addressed important issues difficult to integrate into a traditional BSC without compromising the functional idea proposed masterfully by Kaplan and Norton (e.g. public policy, anticompetitive behavior, corruption, cultural respect to the community or region and others can be included).

The *SBSC* allows making a balance between past- and future-oriented, quantitative and non-quantitative, financial and nonfinancial information (Schaltegger

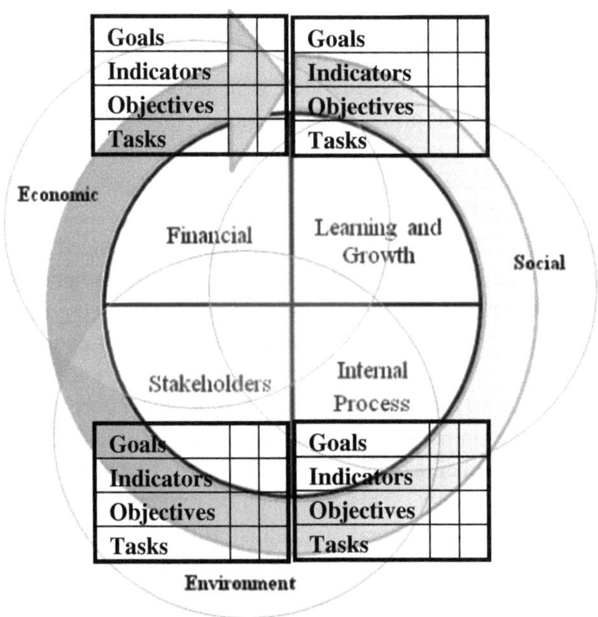

Fig. 1 Sustainability Balanced Scorecard enhanced by stakeholder's perspective [Source: adaptation of (Figge et al. 2002)]

and Dyllick 2002) and include the triple bottom line into the core management of the business. Contemplate the acting of the organization from fives possible perspectives: Learning and Growth, Internal Process, Stakeholders, Financial and Social and Cultural.

Figge et al. (2002) suggest three alternatives to include sustainability issues in the BSC.

1. Integrating social and environmental measures within the existing four quadrants: for example, water use and energy efficiency could fall within internal processes; developing renewable, recyclable resources could be a financial measure or a long-term development target.
2. Developing a separate, but linked, sustainability scorecard, perhaps modeled on the templates that are emerging in corporate sustainability reports: for example, there could be social and environmental quadrants for energy use, waste, community impact, employee well-being and so forth.
3. Adding non-market elements to the scorecard: for example, adding environmental and social measures as separate 'quadrants' or 'spokes on the performance wheel'.

The *SBSC* supports the management processes which are necessary to deal with these challenges. *SBSC* facilitates the development in an active way, of a new dynamic control in organizations impelling the coordination and the complementarities among the different areas of the company and allowing the sustainability strategy of the business. The *SBSC* is considered as a sustainability strategic management system and can be used to manage the CS strategy of the business.

4.3 Analytical Hierarchy Process and Analytic Network Process

The Analytic Hierarchy Process (AHP) is a multicriteria decision technique, which was developed by Saaty (1980). AHP is a tool that combines qualitative and quantitative factors in the selection process and is used to prioritize issues in a complex situation where several factors are involved. This method allows the quantification of the relative priority of each alternative on a scale, which emphasizes the importance of intuitive decision-makers and the consistency of their judgments to make comparisons between the various alternatives. According to San-José Lombera and Cuadrado Rojo (2010) provide a flexible analysis and easy to understand complex problems using a hierarchical structure and provides decision-makers a strong basis for decision-making process.

The AHP compares the criteria as scale or intensity couples preference Saaty which varies from a value of 1 indicates equal preference for both criteria and the value 9 means that a criterion is extremely more important than the other. With the results of pairwise comparison the decision matrix is built. In recent years several investigations have shown preference for certain attributes above or below other when the information provided is not complete. The field of environmental engineering and sustainability have not escaped this preference, some examples of the application of AHP in these areas can be observed (Tao and Hung 2003; Damjan and Glavic 2005; Castellanos-Abella and Van Westen 2007; Gómez-Limón and Riesgo 2008; San-José Lombera and Cuadrado Rojo 2010).

According to Hernández et al. (2010) the AHP is a multicriteria decision method most referenced in the literature over the past 20 years. Others like Hermansa et al. (2008) argue that this has been one of the most used techniques for assessing the weights of environmental indicators using as examples: Indoor Environment Index (Chiang and Lai 2002) and Environmental Friendliness (Puolamaa et al. 1996). Also Saaty (2003) argue that the sustainability indicators weights are generally obtained using the decision method AHP.

Despite the wide acceptance of the AHP in the construction of indices, this gives an unrealistic view of natural phenomena that sometimes tend to be more complex, with a greater number of relationships converting the model into a complex structure.

The Analytic Network Process (ANP), was developed by Saaty in 1996, it provides a tool to deal with decisions without assuming the independence of the elements of different levels and the independence of the elements in different levels. The ANP extended AHP method for problems with dependence and feedback among criteria using the approach of the super-matrix (Saaty 1996). According to Hernández (2010), ANP does not obey the axiom of independence of influence between criteria or alternatives.

The structure of the decision notes that the ANP use the networks without the need to specify levels (Saaty and Saaty 2003). As in the AHP, the domination or influence the relative importance of a central concept, the widely publicized theory multi-criteria AHP is a special case of the ANP.

The ANP is composed of two parts:

1. Control of hierarchy or network objectives and criteria that control the interactions of the system under study.
2. Many subnets of influences between all elements and groups of the problem, one for each control criteria.

The difference between a hierarchy and a network is visible. Hierarchy shows a linear structure from top to bottom without dependency ratios lower to higher levels. The ANP has a network structure that allows the analysis of dependence among elements of the model, which make it more powerful in uncertain situations and let the problem analyzed closer to reality.

ANP is supported by *SuperDecisions* software, developed and coordinated by Saaty, that facilitates the calculation process and it is available on: http://www.SuperDecisions.com.

4.4 Composite Index

Sustainability problems cannot be analyzed or understood if are not considers an integral perspective, they are the results of multiple interacting factors. Scientists are interested in statistically usable data and maybe not in aggregate data, while business managers require aggregate data, which give an idea of goals and criteria fulfillment. Others like the stakeholders prefer rates and it's allow the company do not give operation system details on itself, but expose a picture of their performance.

The index offers decision makers condensed information for performance monitoring, policy progress evaluation, benchmarking comparisons, and decision making (Esty et al. 2005). The indexes are an aggregation of statistics and/or indicators, which often summarized a lot of related information, using an organized method of weighting, scale, and normalization, adding multiple variables into a single summary.

The main objectives of sustainability indexes aggregation are:

- Summarize the existing data related to sustainability issues.
- Communicate information about the sustainability performance.
- Comparability in a period of time.

The literature show a strong tendency to composite index construction a examples can be seen in Puolamaa et al. (1996), Cherchye and Khuosmanen (2002), Chiang and Lai (2002), Damjan and Glavic (2005), Castellanos-Abella and Van Westen (2007), Gómez-Limón and Riesgo (2008), Blanc et al. (2008) all those in environmental and sustainability areas.

In Cuba nowadays can be appreciated lack of mechanisms for environmental and sustainability performance measurement in terms of business objectives, focusing on a single numerical index widely accepted by companies.

Some business approaches have been studied in recent years in Latin America like Ramos and Melo (2006) which make use of questionnaires and evaluations to determine a composite index of environmental performance. The investigation of Broche-Fernández and Ramos-Gómez (2010) is based on an analysis of organizational environmental performance, this include the determination of a comprehensive assessment indicator that takes into account a total of ten variables that are evaluated qualitatively using a numerical equivalent scale, which is recognized as a limiting factor in this proposal.

Another approach can be seen in the research of Sellito et al. (2010) they propose and apply a method for measuring environmental performance. The main objective is capture, with integrated indicators, the complexity involved in environmental systems and how this manifests itself systemically. To do this, divide the environmental impact of the operation, in five subsystems, attributing relative importance and describing the overall impact and process indicators that are evaluated by experts through the Likert scale. Subsequently combine the indicators into a global index which varies between 0 and 100 %.

The main limitation of this model is that it is supports only but expert's judgments as opposed to measures that rely on physical measurements of field variables or mathematical models, which are used as measurements for the calculation of the indicators (Sellito et al. 2010).

For the construction of composite index, are require different steps like: the selection of indicators, homogenization, standardization, weighting and aggregation.

Selection: The decision process of the indicators that comprise the aggregate index.

Homogenization: these step convert the selected indicators of "different nature" to the same criteria either maximize or minimize.

Normalization: the indicators contained in the index, are distributed on different categories, and is needed a common unit or equivalent among them. Some of the most used methods in the literature for standardization or normalization are: Z-score, linear normalization, min–max normalization and others as fuzzy logic.

Weighting: process to determine and assign the relative importance of indicators, based on expert's judgments.

Aggregation: summary of the information in a single value, using mathematical formulas to get the desired index.

5 Research Methodology

According to Lakatos and Marconi (1986), the problem of the research relates to the analysis of a topic or knowledge gap that still has no solution. In this case the scientific problem was identified: *the lack of procedures in Cuba for sustainability assessment*, to integrate consistent indicators related the needs of company management. The sustainability performance evaluation should be supplemented with outcomes measures to determine what's policies, strategies and targets are effective. The research method was mixed that combines qualitative and quantitative survey of data.

To perform the investigation, was first used, a qualitative survey. The collection of data pointed to a theoretical study of performance evaluation process and best practices in business.

The energy issue has been a priority of the Cuban government since the triumph of the Revolution, increasing the generation of 56–94.2 % in the period between 1959 and 1989. In 2004 the energy situation in Cuba was becoming critical by the combination of several factors, and distributed generation turned out to be the big decision to take to resolve the difficult energy situation.

In mid-2004 arise the "Program for energy efficiency in generation," the main goal was implement a strategy that would bring generation to consumption, reducing the dependence of large thermal plants, and promoting increased efficiency and the incorporation of gas plants. This program increased the installed generating capacity in the country by 22 %, an in-crease in fuel consumption of only 4 % and reduction of greenhouse gas emissions in 69 %.

The introduction of DG in Cuba, since 2004, has generated benefits such as increased energy efficiency, decentralized generation, reduced transmission losses and greenhouse gas emissions. Despite these power plants cause a range of negative environmental impacts on the environment as emissions of greenhouse gases, high levels of noise, emissions, liquid waste, among others that are of great interest to stakeholders.

For the application of sustainability performance evaluation procedure, the energy sector was chosen, specially the distributed generation power plants (DGPP) belonging to the province of Villa Clara, Cuba. The DG is defined according Ackermann et al. (2001) as electric power generation units connected directly to the distribution network or connected to the network on the customer site of the meter.

The primary data collection was through interviews with managers to clarify the principal business strategies, group work to identify sustainability aspects and principals impacts; to select the sustainability indicators in order to evaluate de business performance.

The second phase of research was the implementation of the evaluation process through the *sustainability performance evaluation procedure*. These phase supports the quantitative survey, the indicators values collection, identify causal relationships among proposed metrics and the weighting process with ANP and AHP. Finally with the calculated values of *CISP* in each power plant, the comparison process was done. The interpretation and validation of results was accomplished by the experts group, allowing quantify the business sustainability performance and identify critical points in the performance of the studied SPP.

6 Information Technology Supporting Business Sustainability

Information technology (IT) can play an important role in sustainability management, specifically in sustainability performance evaluation. Some examples of the potential of IT include the collection of data on inputs and outputs of different processes, processing and storage of large volumes of data and dissemination of information to different stakeholders (Page and Rautenstauch 2001).

The web applications can facilitate information and data management of sustainability performance evaluation process. The principals' benefits of web applications are:

- Increased accessibility and quality of data.
- Decreased coordination efforts and time optimization.
- Reduced time for manual processing of data from different reports.
- Homogeneous structure of the data.
- Eliminate data redundancy.

In recent years, there have been a different techniques and frameworks to facilitate the development of dynamic web applications, which can play a decisive role in the development of these applications to support the data generated by organizations performance.

In Cuba organizational information related sustainability becomes difficult to collect, the best results are in the field of environmental statistics in government official reports. Business answers to key questions such as: What to measure? How to measure? When measured? left without a clear answer for many organizations, showing difficulties to obtain regular information.

Other problem is the information storage and availability, the lack of information technology support on sustainability performance evaluation. In recent years it has been an important issue, despite in Cuban business sector hasn't been covered properly and inclusiveness found limitations in their research, practical application from the IT perspective.

7 Methodological Contribution

The contribution of this research comes to solve the previously exposed *lack of procedures in Cuba for sustainability performance evaluation*, for that reason was considered the necessity to combine a methodological approach for sustainability assessment, due to the lacks of tools that make operative in Cuban organizations. The procedure proposed is shown in the Fig. 2.

The procedure has an initial phase "*Organization and strategic analysis*", where the study is organized and clarified the principal's strategies of the organization. In a second phase "*Business process inventory*" represent principals processes to identify the inputs and outputs of the processes and identify the main triple bottom line problems associated with the organization. At this stage should be spelled out the main aspects and associated impacts. In phase three "*Sustainability indicators selection*" is made an initial selection by the experts of different indicators based on the significant triple bottom line impacts. Subsequently are distributed these indicators in four perspectives of a Sustainability Balanced Scorecard (Fig. 2), which represent different areas of key organizational results.

In this phase should be documented the indicators selected and defined with the following attributes:

- *Name of indicator*: Describes a synthetic and clear the purpose of the indicator.
- *Type*: in that group will be rated the indicator: economic, environment or social.
- *Process* of the company it is associated with: specify the business processes that relate either directly or indirectly, with the indicator.
- *Person responsible*: employee charged with taking measurements and updates the indicator.
- *Unit of measurement*: units in which the indicator will be expressed.

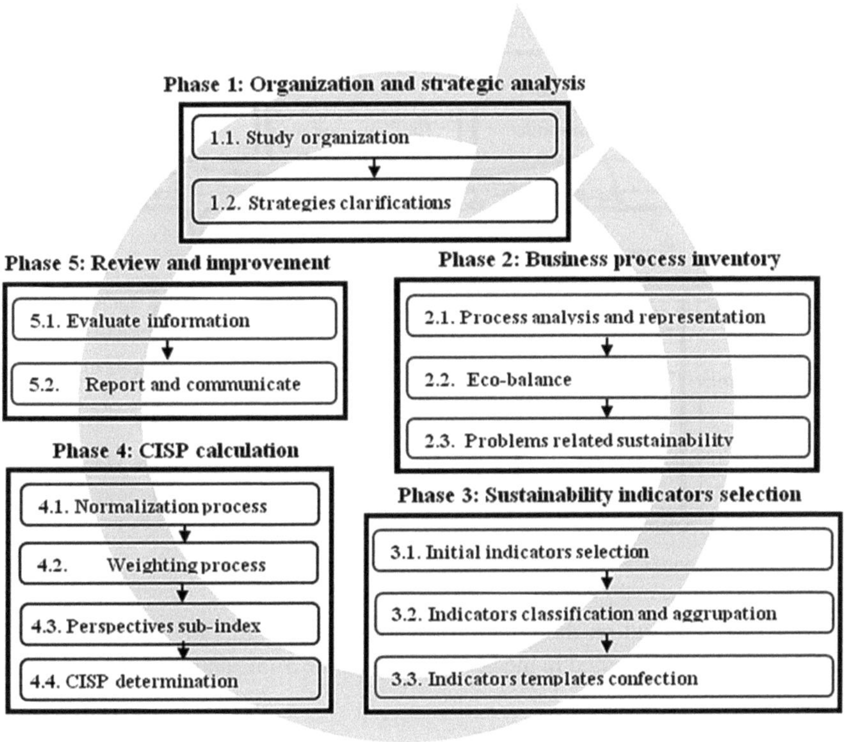

Fig. 2 Procedure for sustainability performance evaluation

- *Frequency of measurement*: indicates the frequency to measure each of the variables involved.
- *Strategic objective*: to which replies will be referred to the business strategic objective of the company or business unit.
- *Calculation Method*: mathematical representation of the indicator.
- *Goals*: targets for the indicator in the short, medium and long term.
- *Relations with other indicators*: should specify relationships with other indicators and the nature of the relationship (direct or indirect).

After defined sustainability indicators and spent the time needed to collect a firs indicators set proceeds to step four.

Phase four, proposed the calculation of *Corporate Index of Sustainability Performance* (see Fig. 3).

The *CISP* is distributed over three clusters: (1) dimensions, perspectives e indicators synthesizing at a rate, the progress or degeneration in corporate sustainability performance, to verify in a simple and continuous way, if the management efforts, administrative management tools and employees training translate into a better or worse business performance.

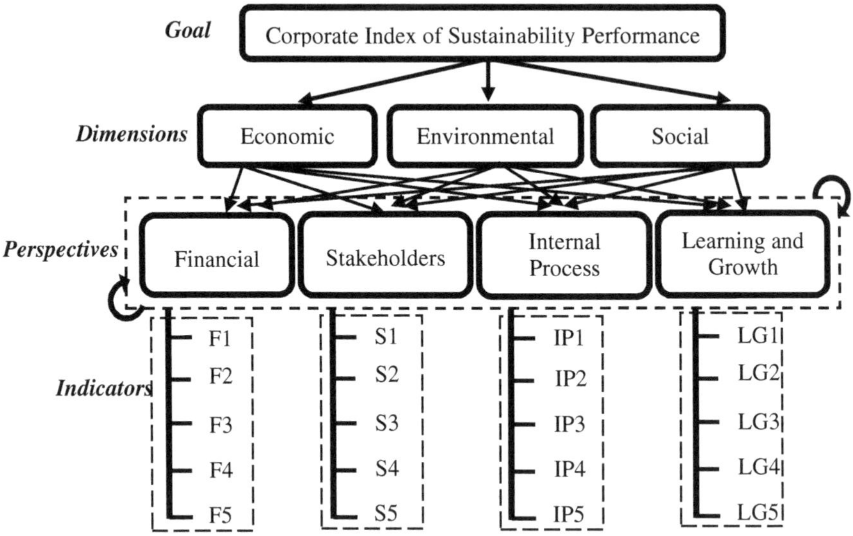

Fig. 3 Corporate index of sustainability performance

The *CISP* is based on the three dimensions of indicators defined in the triple bottom line, and distributed in the perspectives of the *SBSC* and it can be expressed by formula 1.

$$\mathrm{CISP} = \sum_{j=1}^{j=4} \sum_{i=1}^{i=n} Wp_j Wi_{ij} R_{ij} \tag{1}$$

CISP Corporate Index of Sustainability Performance.
Wp_j The relative weight of the perspective j.
Wi_{ij} The relative weight of the indicator i in the perspective j.
Rij Rate or normalized value of the indicator i of the perspective j.

To calculate the index, are determined the weight of each perspective and indicator in each perspective. Weights determination was using multicriteria methods like Analytic Hierarchy Process, Analytic Network Process and the software *SuperDecisions*. The normalization of the indicators can be done through the formula 2.

$$R_{ij} = \begin{cases} \dfrac{x_{ij}}{\max\{x_{ij}\}} & \text{if } x_{ij} \text{ satisfies the condition``more is better''} \\ & 1 \text{ if } x_{ij} \geq \max\{x_{ij}\}\text{``more is better''} \\ \dfrac{\min\{x_{ij}\}}{\{x_{ij}\}} & \text{if } x_{ij} \text{ satisfies the condition ``less is better''} \\ & 1 \text{ if} x_{ij} \leq \min\{x_{ij}\}\text{``less is better''} \end{cases} \tag{2}$$

Rij Rate or normalized value of the indicator i of the perspective j.
x_{ij} value of the indicator to normalize:
i number of the perspective: of 1 at 4.
j number of the indicator: of 1 to *n*.

The procedure defines four sub-indexes (see formula 3) that match with the four perspectives of *SBSC* and they will allow illustrate the indicators behavior in the perspective.

$$PI_j = \sum_{i=1}^{i=n} Wi_{ij}R_{ij} \tag{3}$$

PIj Sub-index of the perspective j.
Wi_{ij} Relative weight of indicators i in the perspective j.
Rij Normalized value of the indicative i of the perspective j.

This sub-indexes allows express the individual performance of each set of indicators in the perspectives of the *SBSC*.

Other concept is introduced in this phase "*improvement potential*" (see formula 4), it has the objective to identify the most influents indicators in relation with the *CISP*.

$$\text{Improvment Potential}_{ij} = Wp_j * Wi_{ij} * \left(1 - R_{ij}\right) \tag{4}$$

In the last phase "*Review and improvement*" the value of the *CISP* is compared versus the scale of sustainability performance evaluation (Table 1).

The main objective of this scale is to provide qualitative meaning to the numerical results of the *CISP*. The preparation of the scale was conducted by experts group with specialists, based on several scenarios of the index and the nine points that divide the Saaty scale for AHP and ANP multicriteria methods in

Table 1 Proposed sustainability performance evaluation scale

Saaty	Range	Evaluation level
9	$0.95 \leq CISP \leq 1$	***Very well:*** the business sustainability performance is adjusted very well to the goals defined in the organizational strategies
7–8	$0.85 \leq CISP < 0.95$	***Well:*** the business sustainability performance is adjusted well to the goals defined with some possibilities of improvement
5–6	$0.75 \leq CISP < 0.85$	***Regular:*** the business sustainability performance is adjusted regular to the sustainability goals and has significant improvements potentials
3–4	$0.65 \leq CISP < 0.75$	***Deficient:*** is deficient with respect to sustainability goals defined by the organization and has several opportunities for improvement
1–2	$0.65 < CISP$	***Poor:*** the business sustainability performance is bad regarding the defined sustainability goals and has large opportunities of improvements

relation to priorities set. The lower limit was set as 0.65 taking into account the normalization method, where the values are ordered separate from the value leader, the goal.

The interpretation must be consistent with the goals and the proposed scope and results. The validity of the data used must be verified by the expert's group.

CISP analysis and its sub-indexes can help to identify the critical points in the sustainability performance, allowing managers refocus organizational efforts towards the worst indicators.

Other important stage in this phase is "*Report and communicate*", this stage intend to provide information and communicate sustainability performance to managers and stakeholders. Also at this stage should be prepared a report with the main results of the procedure application, to help the organization and stakeholders to understand business performance.

8 Application Design

Based on the analysis of the proposed procedure for sustainability performance evaluation, is defined, technology, architecture, and database that support the Web application *System for Sustainability Performance Evaluation* (*SySPE*). The main objective of *SySPE* is support the indicators data acquisition, data storage, aggregation process and graphic representation and report generation. *SySPE* has three main modules (Fig. 4).

The architecture of *SySPE* can be observed in "Fig. 5". To design the *SySPE* database, were considered important elements that should be considered in the design as:

• Sustainability strategies	• Eco-balances	• Process
• Sustainability indicators	• Impacts	• Risks
• Actions	• Perspectives	

The design of *SySPE* is based on the class diagram (Fig. 6).

8.1 Technologies

The technologies used for application development were:

- *MySQL GUI v*8.82: were used the database manager to support the storage of data related to the application.
- *Propel Object Relational Mapping*: eliminates incompatibilities between the relational database language and object-oriented programming. Converting the

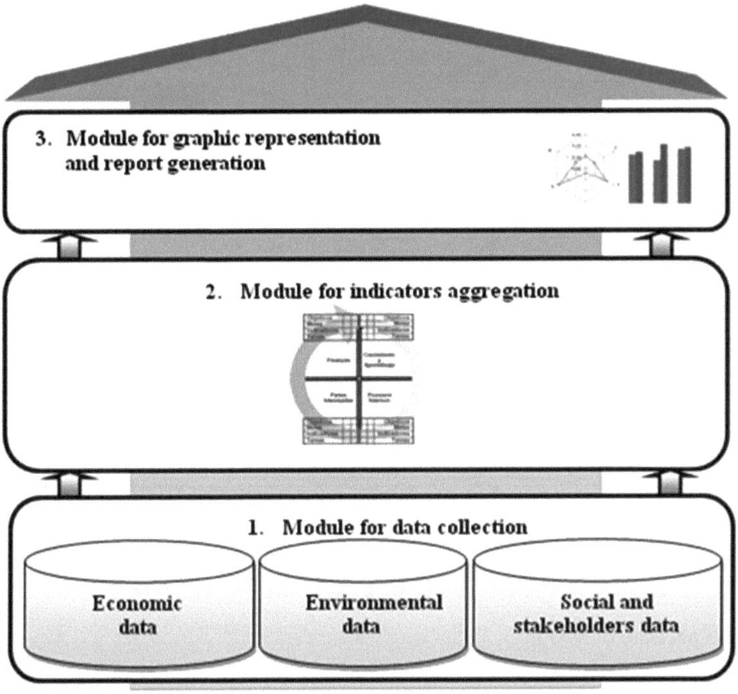

Fig. 4 Modules for *SySPE* application

database schema XML in data objects. Making possible to access and manipulate objects without considering how they are related in correspondence to the data source.

- *Zend Framework*: is responsible for controlling access to the database, implement Model View Controller architectural pattern, achieving modularizes the application, to reuse code and make use of several user interfaces.
- *Eclipse*: is used as an integrated development environment for developing open source application platform *SySPE*. This platform has typically been used to develop integrated development environments.
- *Ext JS*: This is a JavaScript library for developing interactive web applications using technologies such as AJAX, DHTML and DOM. It has a set of components to include in a web application, such as boxes and text areas, fields for dates, numeric fields, combos, HTML editor, toolbar, Windows-style menus and panels divisible into sections.
- *XAMPP*: is used as server platform-independent, free software, which consists mainly of the MySQL database, Apache web server and interpreters for scripting languages: PHP and Perl. The program is licensed under the GNU web server acts as a free, easy to use and capable of interpreting dynamic pages.
- *Business Intelligence and Reporting Tools (BIRT)*: is a project of open source software that provides capabilities for reporting and business intelligence for

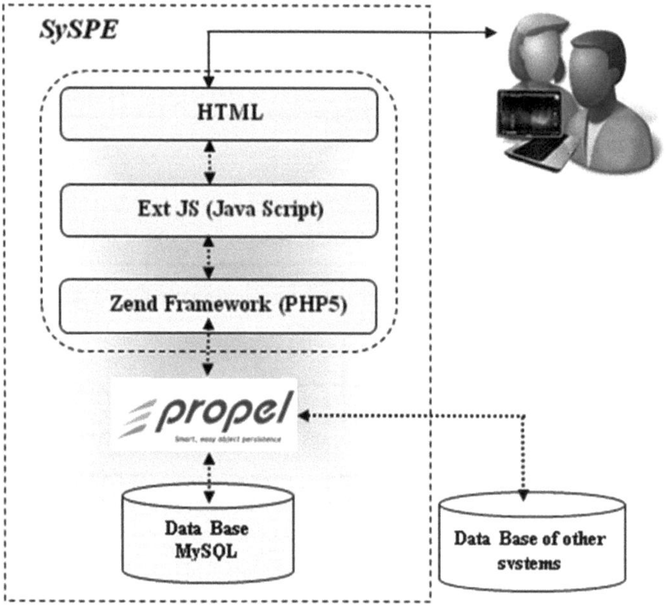

Fig. 5 Architecture of *SySPE*

web applications. *BIRT* also includes a graphics engine that is built into the report designer and can also be used separately to include graphics in an application.

The main window of *SySPE*, can be observed (Fig. 7) with all the principals elements in the menu, that will be handled by the application.

9 Study Case

The study case was carried out in four distributed generation power plants. The first phase set the experts group and serves to define the scope, clarify the sustainability strategies and politicians, to set priorities in the next phases. The second phase helps to characterize and familiarize with the generation process and identify main sustainability aspects and impacts.

Taking like base the previous phases, the strategies, the politicians and impacts where selected initially a total of 27 indicators, contained in triple bottom line dimensions.

These indicators were grouped into the four perspectives of *SBSC* and ordered to assess what should be included in order of importance, limiting the number of indicators selected by five at most, per perspective. Table 2 shows the indicators for each perspective remained as assessed by the expert group.

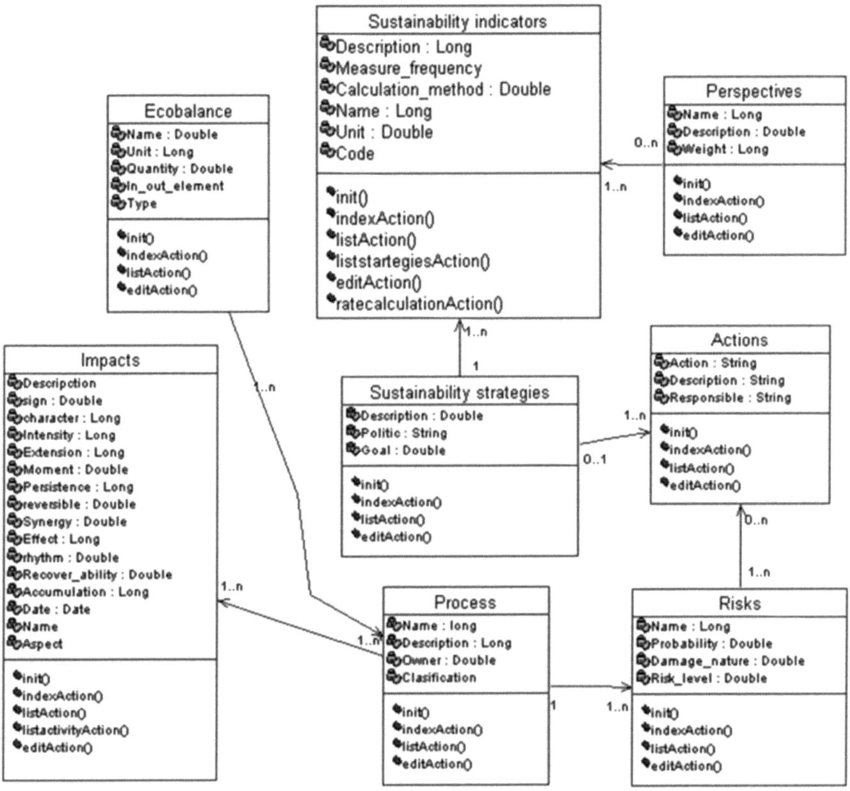

Fig. 6 *SySPE* class diagram

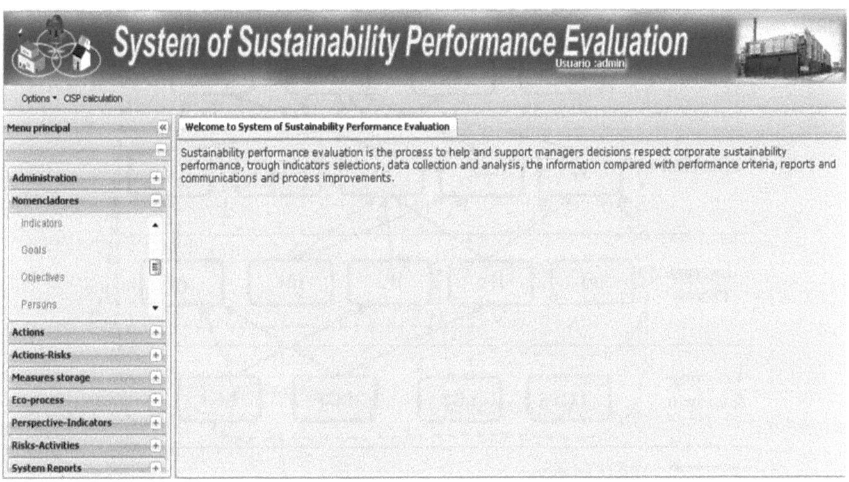

Fig. 7 Main window of SySPE application

Table 2 Final selected indicators by perspectives

Perspectives		Indicators
Financial	F1	Generation cost ($/MW)
	F2	Investment in triple bottom line ($/year)
	F3	Cost related triple bottom line ($/year)
	F4	Fines
Stakeholders	S1	Number of environmental incidents
	S2	Average deficiencies per audit
	S3	Regulatory compliance (%)
	S4	Stakeholders complaints
	S5	Stakeholder's satisfaction (%)
Internal process	IP1	Fuel specific consumption (gr/kWh)
	IP2	Generated muds and residual waters (m^3)
	IP3	Water consumption per kW (m^3/MW)
	IP4	Noise levels (dB)
	IP5	Greenhouse gas emissions (CO_2 e)
Learning and growth	LG1	Number of employees with environmental requirements in the description of their jobs
	LG2	Business sustainability improvement solutions generated by workers
	LG3	Surveys results of employees about their knowledge related sustainability issues in the organization (%)
	LG4	Average hours of training per employee (h/semester)

Were identified relationships among the indicators that finally were selected, the causal relationship map of indicators can be observed in Fig. 8. This map helps experts in indicators weighting process.

For the *CISP* calculation were emitted the experts judgments in the different levels. The first judgments are related to triple bottom line dimensions importance

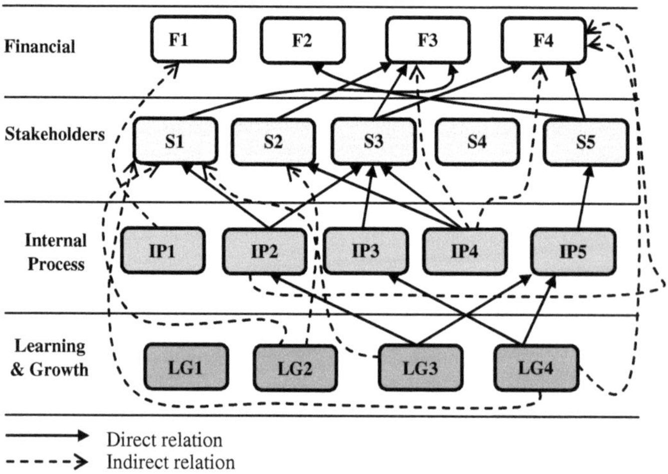

Fig. 8 Indicators causal relationships

Table 3 Dimensions expert's judgments

I = 0.015	Economic	Environmental	Social	Vector
Economic	1	0.33	0.5	0.17
Environmental	3	1	1	0.44
Social	2	1	1	0.39

Table 4 Judgments made about the importance of dimensions on each perspective

		Financial	Stakeholders	I. Process	Learning	Vector
I = 0.09	Financial	1	2	0.25	0.33	0.12
Economic	Stakeholders	0.5	1	0.25	0.25	0.08
	I. Process	4	4	1	4	0.55
	Learning	3	4	0.25	1	0.25
I = 0.05						
Environment	Financial	1	0.33	0.5	2	0.17
	Stakeholders	3	1	2	2	0.42
	I. Process	2	0.5	1	2	0.27
	Learning	0.5	0.5	0.5	1	0.14
I = 0.05						
Social	Financial	1	0.5	0.33	0.33	0.10
	Stakeholders	2	1	0.33	0.33	0.14
	I. Process	3	3	1	3	0.48
	Learning	3	3	0.33	1	0.28

Table 5 Perspectives interactions by expert's judgments

	Financial	Stakeholders	I. Process	Learning	Vector
I = 0.04					
Financial					
Stakeholders		1	3	5	0.64
I. Process		0.33	1	3	0.26
Learning		0.2	0.33	1	0.10
I = 0.01					
Financial	1		2	3	0.55
Stakeholders					
I. Process	0.5		1	1	0.24
Learning	0.33		1	1	0.21
I = 0.04					
Financial	1	0.2		0.167	0.08
Stakeholders	5	1		1	0.44
I. Process					
Learning	6	1		1	0.47
I = 0.04					
Financial	1	3	0.33		0.27
Stakeholders	0.33	1	0.25		0.61
I. Process	3	4	1		0.12
Learning					

Table 6 Internal indicators dependency on each perspective by expert's judgment

0.09	F1	F2	F3	F4	Vector	0.08	S1	S2	S3	S4	S5	Vector
F1	1	2	0.5	3	0,29	S1	1	0.33	0.25	2	0.33	0.09
F2	0.5	1	0.33	0.33	0.11	S2	3	1	0.5	3	0.25	0.17
F3	2	3	1	3	0.43	S3	4	2	1	3	2	0.35
F4	0.33	3	0.33	1	0.17	S4	0.5	0.33	0.33	1	0.33	0.07
						S5	3	4	0.5	3	1	0.31
0.07	LG1	LG2	LG3	LG4	Vector	0.08	IP1	IP2	IP3	IP4	IP5	Vector
LG1	1	3	0.25	0.33	0.14	IP1	1	4	1	7	3	0.38
LG2	0.33	1	0.2	0.25	0.07	IP2	0.25	1	0.33	3	1	0.12
LG3	4	5	1	0.5	0.35	IP3	1	3	1	5	2	0.31
LG4	3	4	2	1	0.44	IP4	0.14	0.33	0.2	1	2	0.08
						IP5	0.33	1	0.5	0.5	1	0.10

Table 7 External indicators dependency by expert's judgment

I = 0		IP3	IP5	Vector	I = 0		S2	S3	Vector
LG4	IP3	1	4	0.8	IP4	S2	1	0.25	0.2
	IP5	0.25	1	0.2		S3	4	1	0.8
I = 0		IP2	IP5	Vector	I = 0		F3	F4	Vector
LG3	IP2	1	3	0.75	IP4	F3	1	0.33	0.25
	IP5	0.33	1	0.25		F4	3	1	0.75
I = 0		S1	S2	Vector	I = 0		S1	S3	Vector
LG2	S1	1	3	0.75	IP2	S1	1	0.2	0.167
	S2	0.33	1	0.25		S3	5	1	0.833
I = 0		F2	F4	Vector	I = 0		F3	F4	Vector
S5	F2	1	0.33	0.25	S3	F3	1	0.25	0.2
	F4	3	1	0.75		F4	4	1	0.8

(Table 3) using ANP. In all the cases should keep in mind the inconsistency (I) of the judgments which must be less than 10 %.

Similarly the judgments were emitted related the influence of three dimensions on each perspective (Table 4), in this case answers the question: How important are dimensions in the different perspectives?

The judgments related perspectives interaction were analyzed and emitted (see Table 5).

The relative importance of indicators on the perspectives, were calculated (see Table 6).

Based in the causal relationships defined (Fig. 8), the importance of dependency among different indicators were emitted by the experts judgments (see Table 7).

Table 8 Finals weights calculated by *SuperDecisions* software

$Wp_F = 0.24$			$Wp_S = 0.29$			$Wp_{IP} = 0.26$			$Wp_{LG} = 0.21$		
Financial	F1	0.01	Stakeholders	S1	0.07	Internal process	IP1	0.015	Learning and growth	LG1	0.143
	F2	0.066		S2	0.002		IP2	0.399		LG2	0.071
	F3	0.214		S3	0.654		IP3	0.361		LG3	0.357
	F4	0.71		S4	0.001		IP4	0.003		LG4	0.429
				S5	0.273		IP5	0.222			

Table 9 Normalized indicators values of four DGPP

		DGPP1	DGPP2	DGPP3	DGPP4
PI_F	F1	0.8	0.99	0.88	1
	F2	0.8	1	1	1
	F3	0.8	0.88	0.9	0.93
	F4	0.71	1	1	1
PI_S	S1	0.71	1	1	1
	S2	0.43	1	0.71	1
	S3	0.89	0.91	0.89	0.91
	S4	1	0.71	1	0.71
	S5	0.75	0.88	0.85	0.94
PI_{IP}	IP1	0.99	1	0.99	0.8
	IP2	0.82	0.86	0.76	0.9
	IP3	0.9	0.85	0.82	0.84
	IP4	0.76	0.81	0.77	0.88
	IP5	0.99	0.99	0.98	0.99
PI_{LG}	LG1	0.81	0.8	1	1
	LG2	0.43	0.71	0.43	0.43
	LG3	0.83	0.93	0.84	0.91
	LG4	0.85	0.93	0.9	1

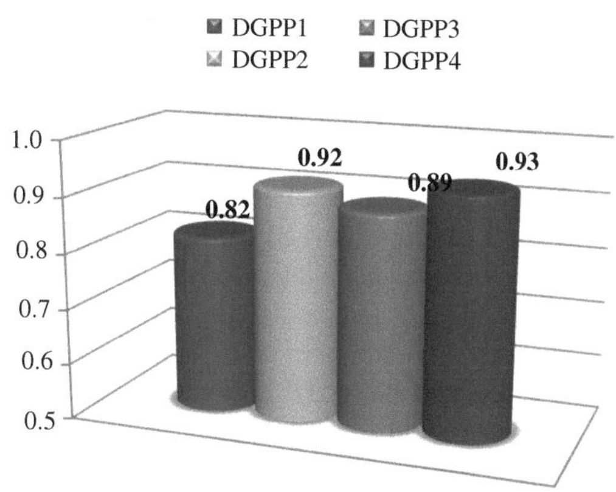

Fig. 9 *CISP* indexes of four DGPP

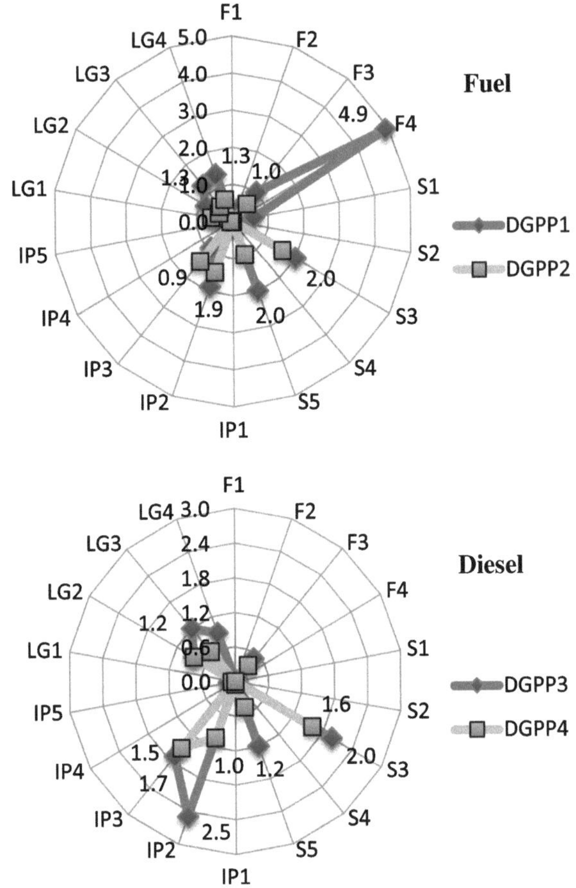

Fig. 10 Improvement potentials of indicators

All these judgments are introduced in *SuperDecision* software to synthetize the final weights trough the weighted super-matrix construction. The weighted super-matrix its form by the local priority vectors been multiplied times the cluster weights. The software gave as results the follow weights (see Table 8).

The first measures of indicators set for the four DGPP normalized by the formula 2 can be observed (see Table 9).

The *CISP* was calculated for the four DGPP using formula 1, with the weights of perspectives e indicators and normalized values (see Fig. 9).

Proceeded to compare and evaluate sustainability performance. Being the first time calculated the *CISP* and being the DGPP similar, comparisons were established among indexes.

Using sustainability performance evaluation scale defined in Table 1, were evaluated: DGPP1 regular, DGPP2, DGPP3 and DGPP4 were evaluated well, although in all cases, a broader scope for improvement sustainability performance exists. For one more deep analysis are introduced indicators improvement potentials (Fig. 10) to identify which indicators affect more the *CISP*.

In DGPP1, the weaknesses identified were the fines (F4), regulatory compliance (S3) and muds and residual waters (IP2). In DGPP2 the main problems are related to S3, P2 and water consumption per kW (IP3).

In relation with DGPP3, indicators of regulatory compliance, water consumption and residual waters have the greatest potential for improvement and DGPP4, the indicators with more improvement potential are: IP2, S3, IP3 and Business sustainability improvement solutions generated by workers (LG2).

10 Conclusions

The proposed procedure allows evaluate the business sustainability performance, establishing a line of action to select, collect, analyze, integrate and evaluate corporate indicators trough the triple bottom line.

The application of the procedure in four DGPP permitted proving its feasibility of implementation as a methodological tool to evaluate sustainability performance and identify critical issues and opportunities for improvement allowing the business to refocus efforts on the major issues.

The CISP facilitate a comprehensive evaluation process of corporate sustainability performance. The calculation of the CISP in four power plants helps to indicate the level of overall compliance related the sustainability goals defined for each indicator, identify indicators of highest importance to the business and make explicit the improvements potentials of each indicators. It also provides a benchmarking among the distributed generation power plants of this research.

The integration of different information technologies for SySPE application design, demonstrating the potential role of information technologies in sustainability performance evaluation. SySPE support the procedure and provides a valuable tool to support the storage, retrieval and integration of different indicators, facilitating the calculation and graphical representation of CISP and improvement potentials. Resolving one of the shortcomings of business sustainability performance evaluation, the support on tools.

References

Ackermann T, Andersson G, Söder L (2001) Distributed generation: a definition. Electr Power Syst Res 57:195–204. doi:10.1016/S0378-7796(01)00101-8

Bieker T (2003) Sustainability management with the balanced scorecard. Switzerland international summer academy on technology studies—corporate sustainability

Blanc I, Friot D, Margni M, Jolliet O (2008) Towards a new index for environmental sustainability based on a DALY weighting approach. Sustain Dev 16:251–260. doi:10.1002/sd.376

Bonacchi M, Rinaldi (2007) DartBoards and Clovers as new tools in sustainability planning and control. Bus Strategy Environ 16:461–473. doi:10.1002/bse.596

Broche-Fernández Y, Ramos-Gómez R (2010) Procedure for management of studies started in Cuban hotel facilities. Retos Turísticos 9(3):13–19

Castellanos-Abella EA, Van Westen CJ (2007) Generation of a landslide risk index map for Cuba using spatial multi-criteria evaluation. Landslides 4:311–325. doi:10.1007/s10346-007-0087-y

Cherchye L, Khuosmanen T (2002) Benchmarking sustainable development: a synthetic meta-index approach. In: Research program on nonparametric methods in economics of production, natural resources, and the environment, Belgium

Chiang CM, Lai CM (2002) A study on the comprehensive indicator of indoor environment assessment for occupants' health in Taiwan. Build Environ 37(4):387–392. doi:10.1016/S0360-1323(01)00034-8

Damjan K, Glavic P (2005) How to compare companies on relevant dimensions of sustainability. Ecol Econ 55:551–563. doi:10.1016/j.ecolecon.2004.12.011

Epstein J, Marie-Josée R (2001) Sustainability in action:identifying and measuring the key performance drivers. Long Range Plan 34:585–604. doi:10.1016/S0024-6301(01)00084-X

Esterhuyse P (2008) The sustainability balanced scorecard: its theory and applications to companies operating within the South African fishing industry. Thesis, faculty of economic and management sciences, University of Stellenbosch

Esty DC, Levy MA, Srebotnjak T, de Sherbin A (2005) Environmental sustainability index: benchmarking national environmental stewardship. Conn Yale center for environmental law and policy, New Haven

Figge F, Hahn T, Schaltegger S, Wagner M (2002) The sustainability balanced scorecard. linking sustainability management to business strategy. Bus Strategy Environ 11(5):269–284. doi:10.1002/bse.339

Gminder C-U (2005) Environmental management with balanced scorecard. In: Chenoweth J, Bird J (eds) A case estudy of Berlin water company. The business of water and sustainable development. Greenleaf Publishing, Germany, p 53

Gómez-Limón JA, Riesgo L (2008) Alternative approaches on constructing a composite indicator to measure agricultural sustainability. In: 107th Seminar January 30 -February 1. Sevilla, Spain, 2008. European Association of Agricultural Economists

Hermansa E, Van den Bosscheb F, Wets G (2008) Combining road safety information in a performance index. Accid Anal Prev 40(4):1337–1344. doi:10.1016/j.aap.2008.02.004

Hernández C (2010) Modelo de gerenciamento da logística reversa integrado às questões estratégicas das organizações. Universidade Estadual Paulista, Faculdade de Engenharia de Guaratinguetá. Tese Doutorado, Guaratinguetá

Hernández C, Silva Marins FA, Medeiros da Rocha P, Rodriguez Duran JA (2010) Using AHP and ANP to evaluate the relation between reverse logistics and corporate performance in Brazilian industry. Braz J Oper Prod Manage 7(2):47–62

Hubbard G (2009) Measuring organizational performance: beyond the triple bottom line. Bus Strategy Environ 19:177–191. doi:0.1002/bse

Kaplan RS, Norton DP (1996) The strategy-focused organization: how balanced scorecard companies thrive in the new business environment. Harvard Business School Press, Boston

Kaplan RS, Norton DP (2000) Having trouble with your strategy?. Then map it. Harvard Business Review, September-October, pp 50–61

Kates RW et al (2001) Sustainability science. Science 292:641–642

Knoepfel I (2001) Dow Jones sustainability group index: a global benchmark for corporate sustainability. Corp Environ Strategy 8(1):6–15. doi:1066-7938/01/$

Lakatos EM, Marconi MA (1986) Metodología científica. Atlas, São Paulo

Page B, Rautenstauch C (2001) Environmental informatics methods, tools and applications in environmental information processing. In: Environmental information systems in industry and public administration, Madegburg, p 3

Perrini F, Tencati A (2006) Sustainability and stakeholder management: the need for new corporate performance evaluation and reporting systems. Bus Strategy Environ 15:296–308. doi:10.1002/bse.538

Puolamaa M, Kaplas M, Reinikainen T (1996) Index of environmental friendliness: a methodological study. Eurostat/Statistics Finland

Purba R et al (2006) A metric for corporate environmental indicators… for small and medium enterprises in the Philippines. Bus Strategy Environ 18:14–31. doi:10.1002/bse.555

Ramos T, Melo J (2006) Developing and implementing an environmental performance index for the Portuguese military. Bus Strategy Environ 15(2):71–86. doi:10.1002/bse.440

Saaty TL (1980) The analytic hierarchy process: planning, priority setting, resource allocation. McGraw-Hill, New York

Saaty TL (1996) Decision making with dependence and feedback. The analytic network process. RWS Publications, Pittsburgh

Saaty RW, Saaty TL (2003) The analytic hierarchy process (ahp) for decision making and the analytic network process (anp) for decision making with dependence and feedback. http://www.superdecisions.com/demos_tutorials.php3 Accessed 13 March 2011

San-José Lombera JT, Cuadrado Rojo J (2010) Industrial building design stage based on a system approach to their environmental sustainability. Constr Build Mater 24:438–447. doi:10.1016/j.conbuildmat.2009.10.019

Schaltegger S, Burritt R (2005) Corporate sustainability. In: The international yearbook of environmental and resource economics. Edward Elgar Publishing, eBook

Schaltegger S, Dyllick T (2002) Sustainability management using the balanced scorecard. Glaber, Wiesbaden Germany

Sellito MA, Borchardt M, Pereira GM (2010) Modelagem para avaliação de desempenho ambiental em operações de manufatura. Gest Prod 17(1):95–109

Statistics Finland (2003) Index of environmental friendliness. http://www.stat.fi/tk/yr/ye22_en.html. Accessed 9 July 2012

Tao C-C, Hung C-C (2003) A comparative approach of the quantitative models for sustainable transportation. J Eastern Asia Soc Transp Stud 4:3329–3344

UNCSD (2012) RIO 2012 Issues Briefs. [Online] UNCSD Secretariat (6.0) URL http://www.uncsd2012.org/index.php?page=view&type=400&nr=218&menu=45 Accessed June 22 2012

Wessman H, Pihkola H (2009) Product sustainability. In: Oittinen P, Saarelma H (eds) Principles, processess and quality. 2nd ed. Helsinki: Print Media. 978-952-5216-33-2

Yüksel I, Dagdeviren M (2010) Using the fuzzy analytic network process (ANP) for balanced scorecard (BSC): a case study for a manufacturing firm. Expert Syst Appl 37:1270–1278. doi:10.1016/j.eswa.2009.06.002

Sustainability Indicators: Development and Application for the Agriculture Sector

Jiří Hřebíček, Soňa Valtinyová, Jan Křen, Michal Hodinka, Oldřich Trenz and Pavel Marada

Abstract Current trends of indicators-based sustainability evaluation (i.e., measurement of environmental, social, economic and governance performance) and corporate sustainability reporting are discussed in the chapter focusing on the agriculture sector. From the perspective of agricultural policy, there are two broad decisions to be made: which indicators to recommend and promote to farmers, and which indicators to collect to assist in agriculture policy-making. We introduce several general approaches for indicators to be collected which will assist in policy-making (European Union, the Organization for Economic Cooperation and Development and Food and Agriculture Organization of the United Nations) in the first part of our chapter and, given the differences in decision-making problems faced by these sets of decision makers. We continue in the second part of chapter with indicators to recommend and promote to farmers in the European Union. Those sets of indicators are likely to differ substantially, potentially with little or no overlap between indicators for regional, national or international decision-makers or for assistance in policy-making. The relationship between environmental and sustainability indicators and corporate sustainability reporting is an important issue; and the possibility of the utilization of information and communication technology and XBRL taxonomy is discussed here.

Keywords Agrosystems · Sustainability assessment · Indicators · Sustainability reporting

---

J. Hřebíček (✉) · M. Hodinka · O. Trenz
Institute of Informatics, Faculty of Business and Economics, Mendel University, Zemědělská 1, 613 00 Brno, Czech Republic
e-mail: hrebicek@iba.muni.cz

S. Valtinyová · J. Křen · P. Marada
Department of Agrosystems and Bioclimatology, Faculty of Agronomy, Mendel University, Zemědělská 1, 613 00 Brno, Czech Republic

M. G. Erechtchoukova et al. (eds.), *Sustainability Appraisal: Quantitative Methods and Mathematical Techniques for Environmental Performance Evaluation*, EcoProduction, DOI: 10.1007/978-3-642-32081-1_4,

1 Introduction

Sustainable development indicators are indicators that measure progress made in sustainable growth and development of organizations, regions and countries or the sector of economic activity. They can provide an early warning, sounding the alarm in time to prevent economic, social and environmental damage. They are also important tools to communicate ideas of sustainable development. Indicators for monitoring progress towards sustainable development are needed in order to assist decision-makers and policy-makers at all levels and to increase focus on sustainable development. From about the seventies of the 20th century it has become obvious that economic development is dependent and limited by natural resources which have been recognised to be exhaustible and damageable, therefore beyond the commonly used economic indicators of well-being, social, environmental and institutional indicators have to be taken into account as well to arrive at a broader, more complete picture of societal development. The need to address these issues in an integrated manner was firstly recognized at the United Nations Conference on Environment and Development (UNCED) in 1992.

A core set of 58 indicators and methodology sheets have been available for all countries to use since 1995. This core set was adopted by the United Nations Commission on Sustainable Development (CSD) at its Third Session in April 1995 (EEA Glossary 2012). The third, revised set of CSD indicators was finalized in 2006 by a group of experts from developing and developed countries and international organizations. The revised edition contains 96 indicators, including a subset of 50 core indicators. The guidelines on indicators (UN DESA 2007) and their detailed methodology sheets are now available as reference for all countries to develop national indicators of sustainable development. However, there is only a small subset of CSD indicators (e.g., *Carbon dioxide emissions; Land degradation; Arable and permanent crop land area; Fertilizer use efficiency; Use of agricultural pesticides; Area under organic farming; Water use intensity by economic activity; Wastewater treatment; Fragmentation of habitats; Vulnerable employment, by sex; Labour productivity and Unit labour costs; Intensity of energy use, total and by economic activity; Generation of waste*) focusing mainly on the agriculture sector. Therefore, CSD indicators do not cover complex issues of "sustainability indicators" in the agriculture sector.

We have to analyse further information sources and the definition of "sustainability indicators" in the research project No P403/11/2085 "Construction of Methods for Multi-factorial Assessment of Company Complex Performance in Selected Sectors" (Hřebíček et al. 2011, 2012; Kocmanová and Dočekalová 2012) and the "Methodology of evaluation of crop production systems sustainability for the conditions in the Czech Republic" (Křen et al. 2011), solved by authors of this chapter.

In recent years, the concept of "sustainability indicators" has become prominent also in agricultural science. The idea is that particular characteristics of resources and agrosystem management are monitored and recorded, with the intention that this information serves as an aid for decision-making by farmers and/ or policy- makers on local, regional level, national or international level. Many sustainability indicators for the agriculture sector have been proposed (Binder et al. 2010; Bockstaller et al. 2009; Felice et al. 2012; Gómez-Limón and Sanchez-Fernandez 2010; Hřebíček et al. 2012; Křen et al. 2011; Pannell and Glenn 2000; Payraudeau and van der Werf 2005; Roth 2010; Singh et al. 2009). Environmental effects of agriculture originate on the level of single farms, therefore tools for optimisation on this level are also needed. The development started in the nineties of the 20th century and there are about 150 methods for farm level sustainability assessment documented today (Rosnoblet et al. 2006). One of the first activities in this issue was done within the Research Network on Integrated and Ecological Arable Farming Systems for EU and associated countries, where 25 research teams from 15 European countries participated (Vereijken 1994). The aim of this project was to provide farmers and advisors in agriculture with tools for continuous improvement of farm performance regarding its environmental effects, resources consumption and sustaining its productivity. In the recent years there are attempts to finalise the assessment process by granting farms a certificate (label) (DLG 2012) to provide them with an advantage for communication with business partners and the public.

The effects of agriculture on the environment are being created by the individual farms, it is therefore necessary to optimize the tools at the level of individual farms.

From the different position of national or international decision-makers are viewed indicators in international forums as essential for the transition towards global sustainable development Moldan et al.(2012). For example we can mention the indicators proposed by the European Union (EU) (CEC 2006, 2009) and the European Environment Agency (EEA) (EEA 2012; Martin et al. 2012; Stanners et al. 2007), furthermore, the Food and Agriculture Organization of the United Nations (FAO) (FAO 1996, 2003, 2012), the Organisation for Economic Cooperation and Development (OECD) (OECD 1997, 2006, 2008a, b, 2010, 2012), and the World Commission on Environment and Development (WCED) in (WCED 1987), etc. The concept of global sustainability apparently has great appeal with regard to environmental and resource management, yet its applicability in practical decision-making is hampered by the ambiguity of its meaning, and the multiplicity of definitions that have been proposed (Pannell and Schilizzi 1999; Pannell and Glenn 2000) and summarized by Moldan et al. (2012). The idea of sustainability indicators seems to have grown out of the recognition that sustainability cannot be condensed to a single simple definition. Its multifaceted nature can be dealt with by monitoring a range of indicators of different types on local or global level for individual farmers or regional, national and international decision-makers.

The identification of the end-user and the definition of practical objectives of the indicators and whole sustainability assessment methods are the basic steps for their development (Binder et al. 2010; Bockstaller et al. 2009; Goodlas et al. 2003; van der Werf and Petit 2002).

Two main groups of indicator's users can be distinguished. Decision and policymakers and administration-needs aggregated indicators which provide them with a complex view (information about the state or development) and are used as the support for development of environmentally just policy for the agricultural sector. This kind of indicator is dealt with in the first part of the chapter.

The second group of users, farmers, practitioners, managers of the farms and agricultural enterprises, has direct impact on the performance of the farm and manages its interaction with the environment. They need fairly detailed information and simple methods on how to determine sustainability indicators, which can be used here for identifying risky points in environmental performance and sustainability of the farms.

Therefore, we introduce several general approaches for indicators to collect to assist in policy-making in the first part of our chapter and give the differences in decision problems faced by these sets for decision-makers. In this part we summarize the sets of indicators officially used by European and international (OECD, FAO) institutions (Hřebíček et al. 2011, 2012) on regional, national and international levels.

We continue in the second part of the chapter with an overview of indicators and methods developed in the Europe designated for sustainability assessment on the farm level (Valtýniová and Křen 2011) which provide links to experience obtained by the application of this approach in the Czech conditions (Křen et al. 2011, 2012).

2 Framework of Sustainability Indicators for the Agriculture Sector on Country and International Level

Let us consider sustainability indicators used to assist in policy-making which focus on companies within the economic activity of the Statistical Classification of Economic Activities in the European Community (NACE) coding (NACE 2011): *A—Agriculture, forestry and fishing* section, where we considered all the subsections *A1.xx* of the subsection *A1—Crop and animal production, hunting and related service activities* excluding subsection A1.07 (hunting, trapping and related service activities).

2.1 EU Agri-environmental Indicators

Before determining the sustainability indicators for the investigated agriculture sector we have to consider and analyse the EU legislation, i.e., the *Common Agricultural Policy* (CAP) after 2013 (CAP 2012) including the reporting needs of other EU policies that relate to sustainability indicators, particularly the *Agri-Environmental Indicators* (EU AEI(s)) and requirements on the collection of related data. The CAP of the EU is due to be reformed by 2013. On 12 October 2011 the European Commission (EC) presented a set of legal proposals designed to make the CAP a more effective policy for a more competitive and sustainable agriculture and vibrant rural areas (CAP 2012).

The development of the EU AEIs is a long-term project for monitoring the integration of environmental concerns into the CAP, at the latest proposed by the EC on 15 September 2006 in COM (2006) 508 final “Development of agri-environmental indicators for monitoring the integration of environmental concerns into the common agricultural policy” (CEC 2006). The EC adopted 28 EU AEIs to assess the interaction between the CAP and the environment. We took into account also the COM (2000) 20 final “Indicators for the integration of environmental concerns into the common agricultural policy”.

In these Communications, 28 Agri-Environmental Indicators are identified (Table 1) according to the Driving forces—Pressures and benefits—State/Impact—Responses (DPSIR) framework of the EEA (EUROSTAT 2010; Stanners et al. 2007). Simply put, the DPSIR framework specifies social and economic developments drive (D) changes that exert pressure (P) on the environment. As a consequence, changes occur in the state (S) of the environment, which leads to impacts (I) on, for example, agriculture systems and ecosystem functioning, human health and the economy. Finally, societal and political responses (R) affect earlier parts of the system directly or indirectly. In Table 1 is the list of EU AEIs connected with domains of the DPSIR framework and each domain is divided into sub-domains with appropriate indicators.

Ideal spatial scales for AEIs reporting have been proposed Vinther et al. (2011) that are specific to individual AEIs, but are kept realistic, often at NUTS 2 level. The Nomenclature of territorial units for statistics (NUTS) classification is a hierarchical system for dividing up the economic territory of the EU for the purpose of the collection, development and harmonisation of EU regional statistics; socio-economic analyses of the regions; framing of EU regional policies (EUROSTAT 2012). A regional approach rather than a national approach is essential to capture the diversity in farming systems in given territory and the environment.

The frequency of data collection is recommended based on the rate of change of the indicator, but also to enable the detection of trend and to satisfy policy requirements without overwhelming the data provider (Vinther et al. 2011).

The set of AEIs have much in common with CAP requirements, and even those that are not required directly for the CAP are useful for monitoring the outcomes of the CAP implementation. It is therefore expected that the use of the AEIs will provide much needed coordination for data collection at an EU level to meet the needs of the key agri-environmental policies.

In the context of the Renewed EU Sustainable Development Strategy (CEC 2009) of EU, these EU AEIs serve to:

Table 1 List of the EU agri-environmental indicators. Modified from Vinther et al. (2011)

Domain DPSIR	Sub-domain	Indicator code	Title	Unit
Driving forces	Input use	AEI5	Mineral fertiliser consumption	kg/ha
		AEI6	Consumption of pesticides	kg/ha of each substances
		AEI7	Irrigation	% of UAA
		AEI8	Energy use	GJ/ha/year
	Land use	AEI9	Land use change	% of UAA
		AEI10.1	Cropping patterns	% of UAA
		AEI10.2	Livestock patterns	LU/UAA
	Farm management	AEI11.1	Soil cover	% of the year
		AEI11.2	Tillage practices	% of arable area
		AEI11.3	Manure storage	% of safe storage
	Trends	AEI12	Intensification/extensification	€/ha
		AEI13	Specialisation	% of UAA
		AEI14	Risk of land abandonment	% loss of UAA
Pressures and benefits	Pollution	AEI15	Gross nitrogen balance	kg N/ha/year P/ha/year
		AEI16	Risk of pollution by phosphorus	kg P/ha/year
		AEI17	Pesticide risk	Index of risk[a]
		AEI18	Ammonia emissions	ktonnes/year
		AEI19	Greenhouse gas emissions	ktonnes CO_{2eqv}/year
	Resource depletion	AEI20	Water abstraction	m^3/year
		AEI21	Soil erosion	% of UAA
		AEI22	Genetic diversity	Not defined yet
	Benefits	AEI23	High Nature Value farmland	% of UAA
		AEI24	Renewable energy production	% of total energy production
State/Impact	Biodiversity and habitats	AEI25	Population trends of farmland birds	Farmland bird population index
	Natural resources	AEI26	Soil quality	Soil quality index
		AEI27.1	Water quality—Nitrate pollution	mg/l NO_3
		AEI27.2	Water quality—Pesticide pollution	μg/l
	Landscape	AEI28	Landscape—state and diversity	Three supporting indicators
Responses	Public policy	AEI1	Agri-environmental commitments	AE commitments/UAA
		AEI2	Agricultural areas under Natura 2000	% of UAA
	Technology and skills	AEI3	Farmers' training level and use of environmental farm advisory services	Number/year
	Market signals and attitudes	AEI4	Area under organic farming	% of UAA

UAA utilised agricultural area; *LU* livestock units; [a] Kruijne et al. (2011)

- provide information on the farmed environment;
- track the impact of agriculture on the environment;
- assess the impact of agricultural and environmental policies on environmental management of farms;
- inform agricultural and environmental policy decisions;
- illustrate agri-environmental relationships to the broader public.

The last list of EU AEIs collected by Eurostat from Member States (MS) of the EU is summarized at the Eurostat web page (EUROSTAT 2010). It is devoted to track the integration of environmental concerns into the CAP at the EU, national and regional levels. However, the level of the development of these indicators with respect to sustainability differs. Some EU AEIs are already operational, their concepts and measurement are well-defined and data are available at national and, where appropriate, at regional level of decision or policy making. Other indicators are well-defined but they lack regional or harmonised data or their modelling approaches are weak. There are also indicators that still need substantial improvements in order to become fully operational (Hřebíček et al. 2012). Therefore, not all indicators can be disseminated for the time being and we took this into account.

From where can data be obtained to calculate EU AEIs for Member States? What follows are three basic surveys of EU as existing data sources of EU Member States for several of the EU AEIs (Oenema et al. 2011):

- *Farm Structure Survey* (FSS). The Farm Structure Survey (FSS 2012) at Eurostat provides comparable statistical data on the structures of agricultural and horticultural enterprises in all Member States of the EU, e.g., data from the Czech Republic (Martins 2008). These statistics contain data on the number of farms, production sector, form of ownership, land use, crop production, livestock production, farmers and other labour forces on farms, working hours spent on agricultural work, working outside the farm, secondary business activities on farms, organic production, machinery and equipment on farms, manure pits, and irrigated areas.
- *Survey on Agricultural Production Methods* (SAPM). The Survey on Agricultural Production Methods (Charlier 2010) is a one-off supplement to the FSS focusing on production methods and management, and includes questions on the following topics: tillage methods, soil conservation, actions against erosion and nutrient leaching, landscape features, animal grazing, animal housing, nutrients, manure storage and treatment facilities, plant protection, and irrigation.
- *Farm Accountancy Data Network* (FADN). The Farm Accountancy Data Network (FADN 2010) is an instrument for evaluating the income of agricultural holdings and the impacts of the CAP. The concept of the FADN was launched in 1965, when the Regulation No 79/65/EEC of the Council of 15 June 1965 set up a network for the collection of accountancy data on the incomes and business operation of agricultural holdings in the European Economic Community (CEC 1995) and established the legal basis for the organisation of this network. It consists of an annual survey carried out by the Member States of the EU

accountancy data from a sample of the agricultural holdings in the EU every year. Derived from national surveys, the FADN is the only source of micro-economic data that is harmonised, i.e., the bookkeeping principles are the same in all countries. Holdings are selected to take part in the survey on the basis of sampling plans established at the level of each region in the EU. For instance, *Liaison Agency FADN CZ* (FADNCZ 2011) is responsible for carrying out the FADN survey by collecting accountancy data from a sample of the agricultural holdings in the Czech Republic every year. The concept of the FADN in the Czech Republic was launched in 1995 and the Liaison Agency FADN CZ (CLA) was established by the decision of the Minister of Agriculture in 2003. The CLA is guided by a National FADN Committee and coordinates the collection of the data from the surveyed farms in the Czech Republic. Data, consisting of production, structural and financial variables, are collected in a questionnaire for each holding. Currently, the annual sample covers approximately 1.700 holdings. At present, the CLA operates the FADN fulfilling all the functions in relation to the EU. The CLA provides FADN data to experts either in the form of publication or through electronic media. Furthermore, the CLA takes part in the national and international projects in the field of economics of the agricultural sector, information systems and agricultural statistics. The statistical methods are used to test the representativeness of the FADN CZ sample and to analyse data validity.

We can obtain time series of indicators for EU Member States from Eurostat that collects data and calculate chosen AEIs. For example, the AEI15 Gross nitrogen balance of EU Member States indicator is free to download at the Eurostat database (EUROSTAT 2009). We show extracted times series of the Czech Republic from EUROSTAT (2009) in Table 2.

We have also been inspired by the specific recommendations of the "Direct and indirect data needs linked to the farms for agri-environmental indicators" (Dire-Data) project. EU AEIs were introduced in the series of publications (Amon et al. 2011; Oenema et al. 2011; Van Beek et al. 2011; Velthof 2011; Vinther et al. 2011; Wilson et al. 2011) of Eurostat Methodologies & Working papers edited by Selenius, Baudouin and Kremer from Eurostat.

The DireDate project was focused on analysing the direct and indirect data needs linked to the farms, with the objective of setting up an efficient and sustainable data collection for agri-environmental indicators and policy reporting in the EU. The general objective of DireDate project was: "to create a framework for setting up a sustainable system for collecting a set of data from farmers and other

Table 2 Gross nutrient balance per hectare (arable land, permanent crops, permanent grassland) for the Czech Republic. Modified from EUROSTAT (2009)

Years	2000	2001	2002	2003	2004	2005	2006	2007	2008	2009
AEI15 (kg N/ha/year)	72	77	84	82	72	77	88	91	86	60

sources that would serve primarily European and national statisticians for creating the agreed 28 agri-environmental indicators (EU AEIs) and thus serve policy makers, but also agricultural and environmental researchers, observers of climate change and other environmental issues linked to agriculture" (Van Beek et al. 2011; Vinther et al. 2011).

We have used their recommendations about the collection of agri-environmental data and information in the project No P403/11/2085.

2.2 EEA Environmental Indicators for Agriculture

Over the past two decades, the EEA has published assessments and indicators on most European environmental issues. Today it maintains an extensive set of over 200 environmental indicators (EEA Indicators 2012) across 12 environmental themes (Martin and Henrichs et al. 2012), which are: Agriculture; Air pollution; Biodiversity; Climate change; Energy; Transport; Waste; Water; Fisheries; Land and soil; Tourism; Environmental scenarios. EEA indicators are developed for Agriculture against the DPSIR assessment framework (Stanners et al. 2007).

The EEA has developed a formidable store cupboard of environmental, economic and social data and indicators that could be used to a much greater degree than hitherto, to support current EU policy priorities such as the Europe 2020 strategy (EC 2010) and the forthcoming 7th Environmental Action Programme. The EEA aims to deliver timely, targeted, relevant and reliable information to policy makers and the public—environmental indicators play a key role in this.

We have considered the subset of the Core Set of Indicators (CSI) and related indicators published by the EEA (EEA Indicators 2012) focusing on the Agriculture sector (Table 3) as sustainability indicators for the environment. They are ordered in a similar fashion like data in Table 1, in accord with the DPSIR framework.

We can find detailed factsheets and methods of indicator data collection in Table 2 on the web of EEA Indicators (2012) and we do not introduce them here.

2.3 OECD Agri-environmental Indicators

Agriculture also has significant impacts on the environment in OECD countries. The impacts occur on and off farm, including both pollution and degradation of soil, water and air, as well as the provision of ecological goods and services, such as biodiversity and providing a sink for greenhouse gases. To help improve measurement of the environmental performance of agriculture, the OECD has established a set of agri-environmental indicators (OECD 1997, 2008a) while developing the indicators in cooperation with Eurostat and FAO.

The set of OECD agri-environmental indicators (OECD AEIs) has been developed through several specific theme focused workshops involving OECD

Table 3 List of the EEA environmental indicators for agriculture. Modified from EEA Indicators (2012) factsheets

Domain DPSIR	Sub-domain	Indicator code	Title	Unit
Driving forces	Farm management	CSI033	Aquaculture production	tonnes/km
		YIR01AG09	CAP expenditures	Mio€
Pressures	Pollution	CSI025	Gross nutrient balance	kg/ha/year
		CSI002	Emissions of ozone precursors	ktonnes
		APE003	Ammonia (NH_3) emissions	ktonnes
		SEBI016	Freshwater quality	mg NO_3/l or mg(NO_3-N)/l and mg P/l
	Resource depletion	FAO010	Total fertiliser consumption	Megatonnes or kg/ha of arable land
		EEA023	Fertilizer consumption	kg/ha or %
		CSI018	Use of freshwater resources	% of WEI
State/ Impact	Natural resources	CSI 020	Nutrients in freshwater	mg NO_3/l
		WHS 01a	Pesticides in groundwater	μg/l
Response	Market signals and attitudes	CSI026	Area under organic farming	% of UAA
	Technology and skills	YIR01AG11	Agri-environmental management contracts	% of UAA

UAA utilised agricultural area; *WEI* water exploitation index

country analysts and scientific experts, complemented with thorough reviews of the literature (OECD 1997, 2008a, b, c, 2010). The publication OECD (2008a) is the fourth volume of the series OECD AEIs which examines performance across OECD countries in terms of environmental themes: Soil; Water; Air; Biodiversity; Farm management and Agricultural inputs. The OECD's Driving Force-State-Response (DSR) framework was the organising framework for the long time process of developing the OECD AEIs instead of the EU DPSIR analytical framework. Detailed information (the list of OECD AEIs, data sets of OECD countries and publications) is presented on the web of OECD (2012).

The OECD fosters sustainable development by using indicators to analyse and measure the effects on the environment of domestic agricultural and agri-environmental policies and trade measures. OECD trends of environmental and sustainability conditions related to agriculture are examined across nine themes since 1990 (OECD 2008a):

- Agricultural production and land.
- Nutrients (nitrogen and phosphorus balances).
- Pesticides (use and risks).
- Energy (direct on-farm consumption).
- Soil (water and wind erosion).
- Water (use and quality).

- Air (ammonia, methyl bromide use, greenhouse gas emissions).
- Biodiversity (genetic, wild species and ecosystem diversity).
- Farm management (nutrients, pesticides, soil, water, biodiversity, organic).
- Socio-economic (production, structure, employment, support).

The times series primary database of the OECD AEIs is available on the web OECD (2008c) along with the OECD publication "Environmental Performance of Agriculture in OECD Countries Since 1990" (OECD 2008a), where the assessment of these indicators according to OECD indicator criteria—policy relevance, analytical soundness, measurability, and ease of interpretation—is presented. This database provides OECD cross-country coverage on an annual basis since 1990.

We introduce the set of indicators in the Table 4 in a wider context of OECD (2008a), where the sections of this publication in brackets with appropriate indicators are shown. The OECD AEIs are summarized as follows:

- *driving forces* of agricultural production (Section 1.1.2) and land use (Section 1.1.3) are considered as they relate to other *key driving forces*, especially purchased farm input use nutrients (Section 1.2), pesticides (Section 1.3), energy (Section 1.4) and water use (Section 1.6.1), which play a major role in affecting the state of the environment at agriculture sector;
- the *state* of the environment is related to agriculture, both on and off farm soil erosion (Section 1.5), water quality (Section 1.6.2), ammonia (Section 1.7.1), methyl bromide (Section 1.7.2), greenhouse gas emissions (Section 1.7.3), genetic diversity (Section 1.8.1), wild species diversity (Section 1.8.2), ecosystem diversity (Section 1.8.3), and which in turn lead to;
- a *response* by farmers in terms of altering their farming practices and systems nutrient management (Section 1.9.2), pest management (Sect. 1.9.3), soil management (Section 1.9.4), water management (Section 1.9.5), biodiversity management (Section 1.9.6) and organic management (Section 1.9.7).

All of above OECD AEIs are those which relate to agri-environmental issues faced by most OECD countries, and are based on the best available science and data available for a representative group of countries, as shown in Chap. 1 of (OECD 2008a).

Other OECD AEIs for which either methodologies and/or data sets are not yet at a stage that allows for representative comparative OECD country coverage are summarised in the Sect. 2 of (OECD 2008a).

The forth-coming OECD working plan will reduce the number of OECD AEIs from Table 3 to the following 13 OECD AEIs (FAO 2010):

1. Quantity and share of agricultural water use in total national water utilization.
2. Quantity and share of agricultural ammonia (NH_3) emissions in national total ammonia emissions.
3. Quantity of methyl bromide use in terms of tonnes of ozone depleting substance equivalents.
4. Quantity and share of agricultural greenhouse gas emissions in national total greenhouse gas emissions.

Table 4 List of the OECD agri-environmental indicators. Modified from (OECD 2008a)

Domain	Section	Indicator description	Unit
Driving forces	Agricultural production (1.1.2)	National agricultural production	Mio US$
	Land use (1.1.3)	National total land area	kha
		Total agricultural land area	kha of UAA
		Other categories of agricultural land	kha of UAA
	Nutrients (1.2)	33. Gross balance of nitrogen inputs	kg N/ha/year
		34. Gross balance of phosphorus inputs	kg P/ha/year
	Pesticides (1.3)	35. Pesticide use	tonnes/year
		36. Pesticide risk	Index of risk
	Energy (1.4)	37. Farm energy consumption	ktonnes oil equivalent
	Water use (1.6.1)	3. Quantity of water use	Mio m^3
		4. Quantity of groundwater use	Mio m^3
		5. Area of irrigated land	% of UAA
State	Soil erosion (1.5)	1. Area of water erosion	% of UAA
		2. Area of wind erosion	% of UAA
	Water quality (1.6.2)	6. Nitrate and phosphate water pollution	%
		7. Nitrates concentrations in water	mg/l NO_3
		8. Pesticide concentrations in water	μg/l
		9. Pesticides concentration in water	%
	Ammonia (1.7.1)	10. Quantity of NH3 emissions.	ktonnes/year
	Methyl bromide (1.7.2)	11. Quantity of methyl bromide use	ktonnes/year
	Greenhouse gas emissions (1.7.3)	12. Gross total agricultural GHG emissions	ktonnes CO_2eqv/year
	Genetic diversity (1.8.1)	13. Plant varieties marketed	Number
		14. Dominant crop varieties	Number
		15. Area of transgenic crops	ha
		16. Livestock breeds marketed	Number
		17. Dominant livestock breeds	Number
		18. Number of livestock endangered	Number
		19. Status of genetic resources	Number
	Wild species diversity (1.8.2)	20. Wild species using farmland	Number
		21. Populations of breeding birds	Number
	Ecosystem diversity (1.8.3)	22. Conversion of agricultural land	ha
		Area of semi-natural habitats	ha
		24. National bird habitat areas	ha

(continued)

Table 4 (continued)

Domain	Section	Indicator description	Unit
Response	Nutrient management (1.9.2)	25. Nutrient plans	ha
		26. Soil nutrient testing	ha
	Pest management (1.9.3)	27. Integrated pest management.	ha
	Soil management (1.9.4)	28. Soil conservation	ha
		29. Vegetative cover	ha
	Water management (1.9.5)	30. Irrigation technologies	ha
	Biodiversity management (1.9.6)	31. Biodiversity management plans.	ha
	Organic management (1.9.7)	32. Organic farming	ha

UAA utilised agricultural area.

5. Populations of a selected group of breeding bird species those are dependent on agricultural land for nesting or breeding.
6. Gross balance of the quantity of nitrogen inputs (e.g., fertilisers, manure) into and outputs (e.g., crops, pasture) from farming per hectare of agricultural land.
7. Gross balance of the quantity of phosphate inputs (e.g., fertilisers, manure) into and outputs (e.g., crops, pasture) from farming per hectare of agricultural land.
8. Quantity of pesticide use (or sales) in terms of active ingredients.
9. Quantity and share of direct on-farm energy consumption in national total energy consumption.
10. Area and share of total agricultural land in total national land area.
11. Area and share of the main agricultural land use types (i.e., arable crops, permanent crops and pasture) in total agricultural land.
12. Area and share of land under organic farming in total agricultural land.
13. Area and share of land under transgenic crops in total agricultural land.

The complex information about OECD AEIs is available at the new OECD web page of agri-environmental indicators (OECD 2012), where it is possible to find the link to the complete database of OECD AEIs, as well as links to abovementioned themes or to indicators and statistics of OECD countries.

2.4 FAO Agri-environmental Indicators

The FAO is working closely with the OECD and Eurostat in the development, convergence and production of agri-environmental statistics and indicators (FAO 2010, 2011; FAOSTAT 2012). In the view of the growing demand for information about agri-environmental statistics and for the construction of related FAO agri-environmental indicators (FAO AEIs), a variety of domains is being monitored: *Air and climate; Land; Fertilizers; Pesticides; Livestock; Soil; Water*; and *Energy*.

It is organised into the *Driving forces—Pressure—State—Impact—Responses* (DPSIR) framework, which includes 17 indicators, the statistics of which are described by 59 data series (FAO 2003).

Data and derived FAO AEIs come from FAOSTAT (FAOSTAT 2012), other FAO databases, as well as from other international organizations according to the different fields, land and water management, land tenure issues, biodiversity and genetic resources for food and agriculture, research and extension in order to strengthen FAO's role in providing clear directives to countries on national and international issues related to the environment and sustainable development of agriculture, forestry and fisheries.

The definition of the FAO AEIs DPSIR framework and the coordination with the existing OECD and EU frameworks presented above had to face challenges arising from limited data availability. While OECD and EU designed their frameworks primarily over a limited number of Developed Countries (OECD countries and EU Member States), with established and solid structures in the charge of data collection, in the case of FAO the area of interest covers virtually all the globe, and there are great differences in the amount of data available in the different countries and regions and its quality.

The FAOSTAT Agri-environmental domain under development includes 19 indicators (Table 5), described by 68 data series in Annex 1 (FAO 2011), where the FAO AEIs domain in FAOSTAT and corresponding OECD/EUROSTAT frameworks is described.

FAOSTAT, the FAO statistical database, provides time-series and cross-sectional data relating to hunger, food and agriculture for approximately 245 countries and 35 regional areas covering agriculture, nutrition, fisheries, forestry and food aid from 1961 through the present. It also offers an innovative tool, FAOSTAT Analysis, for basic statistical analysis of the data. It provides users with: a comprehensive global collection of statistics on agriculture; comparative analysis tools of sectors, commodity, country and regions; easy download and visualization of data and indicators (FAOSTAT 2012).

Agri-environmental statistics are compiled by FAO through a variety of methods such as censuses, surveys, remote sensing, administrative records, questionnaires and monitoring and network facilities for the development of environmental indicators and assessments. The data published in FAOSTAT (2012) on land, labour, water, fertilizers, pesticides etc., are compiled through the FAO agriculture resource questionnaires, and are intended to serve building resources accounts, and are vital for environmental analysis. The data collected, however, are not adequate for directly assessing the environmental impact of social and economic activities. In addition, the data reflect national averages and aggregates, while environmental problems are, in most cases, site and time specific.

Table 5 List of FAO agri-environmental indicators. Modified from (FAO 2011)

Domain	Subdomain	Indicator	Unit
Driving forces	Fertilizers	Min. Fertilizers Consumption	kg/ha
	Pesticides	Pesticide consumption	kg/ha of each substances
	Water	Area equipped for irrigation	% of UAA
	Energy	Energy use	GJ/ha/year
	Land	Agricultural land use change	% of UAA
		Share agricultural land	% of UAA
		Cropping patterns	% of UAA
		Livestock patterns	LU/UAA
		Conservation agriculture	% of cultivated area
	Air and Climate Change	Ammonia emissions	ktonnes/year
Pressures	Nutrients	Gross nitrogen balance	kg N/ha/year
		Gross phosphate balance	kg P/ha/year
	Air and Climate change	GHG emissions from agriculture	ktonnes CO_2eqv/year
	Water	Water use in agriculture	m^3/year
	Soil	Soil erosion	% of UAA
	Energy	Biofuels	% of total energy production
State/Impact	Soil	Soil quality	Soil quality index
Responses	Land	Agri-environmental commitments	AE commitments/UAA
		Organic agriculture	% of UAA

UAA utilised agricultural area; *LU* livestock units

2.5 FAO View on Sustainability Indicators and Assessment

The FAO developed the *Sustainability Assessment of Food and Agriculture systems* (SAFA) guidelines (FAO 2012) in the same spirit of codes of practice, guidelines and other recommended measures to assist the achievement of fair practices in food and agriculture production and trade on a local and regional level. The SAFA guidelines are the result of an iterative process, built on the *cross-comparisons of codes of practice, corporate reporting, standards, indicators* and *other technical protocols* currently used by food and other companies and organizations that implement sustainability tools.

The structure and methodology of the SAFA Guidelines draw specifically upon: ISO 14040:2006, the ISEAL Code of Good Practice (ISEAL Alliance 2010), the Reference Tools of the Global Social Compliance Programme (GSCP 2010), the Sustainability Reporting Guidelines (G3.1 2011) and its Food Sector Supplement (GRI 2011) of the Global Reporting Initiative (GRI 2006). The SAFA Guidelines will be revised and finalised in 2013 in order to improve their practicality, applicability, usefulness and soundness.

The guiding vision of SAFA is that food and agriculture systems worldwide are characterised by environmental integrity, economic resilience, social well-being and good governance.

Recent years have seen some progress in the realisation of a socially, economically and environmentally sustainable development (Hřebíček et al. 2012; Kocmanová and Dočekalová 2012; Soukopová and Bakoš 2010). Many stakeholders in the agriculture sector have contributed to this progress by improving agricultural productivity, protecting human and natural resources and conceiving and implementing frameworks, standards and indicators for assessing and improving sustainability across the agricultural sector and along the value chain (Valtýniová and Křen 2011).

The SAFA sustainability rating pertains to the four dimensions of sustainability: Good governance (G); Environmental integrity (E); Economic resilience (C) and Social well-being (S), see Table 6. Within these dimensions, 20 sustainability

Table 6 SAFA sustainability dimensions. Modified (FAO 2012)

Core sustainability themes	Sub-themes
G1 governance structure	Corporate ethics; due diligence
G2 accountability	Holistic audits; responsibility
G3 participation	Stakeholder dialogue; grievance procedures; conflict resolution
G4 rule of law	Commitment to fairness and legitimacy; remedy, restoration and prevention; co-responsibility; resource appropriation
G5 holistic management	Sustainability in quality management; certified production and sourcing; full-cost accounting
E1 atmosphere	Greenhouse gases; air pollution
E2 freshwater	Water quantity; water quality
E3 land	Organic matter; physical structure; chemical quality; land degradation and desertification
E4 biodiversity	Habitat diversity and connectivity; ecosystem integrity; wild biodiversity; agri-cultural biodiversity; threatened species
E5 materials and energy	Non-renewable resources; energy supply; eco-efficiency; waste disposal
E6 animal welfare	Freedom from stress; species-appropriate conditions
C1 investment	Internal investment; community investment; long-ranging investment
C2 vulnerability	Stability of supply; stability of marketing; liquidity and insurance; employment; stability of production
C3 product safety and quality	Product information; traceability; food safety; food quality
C4 local economy	Value creation; local procurement
S1 decent livelihood	Wage level; capacity building
S2 labour rights	Employment; forced labour; child labour; freedom of association and bargaining; working hours
S3 equity	Non-discrimination; gender equality; support to vulnerable people
S4 human health and safety	Physical and psycho-social health; health resilience

themes were identified, each of which contains sub-themes based on (UN DESA 2007) which can be used at local (farmers or companies) or regional, national and international (decision-makers, policy-makers etc.) levels. Details on dimensions, themes, sub-themes and indicators are provided in part C of the SAFA Guidelines (FAO 2012).

For a detailed description of the SAFA assessment procedure, see part B of the SAFA Guidelines. To conduct a SAFA, the following phases of sustainability assessment must be run through (FAO 2012):

1. Setting goal and scope of the assessment.
2. Adapting the SAFA Guidelines: relevance and compliance check.
3. Selecting tools and indicators.
4. Collecting data.
5. Analysing and interpreting SAFA results.
6. Reporting.

3 Development and Use of Sustainability Indicators on Farm Level in the Europe

To assess sustainability of agriculture on the farm level, indicators dealing with fundamental features of the agriculture system (agrosystem) have to be included. The agrosystem is a production system absolutely dependent on basic biological processes which distinguishes it from other sectors of economic activities. Its main feature, soil fertility, is defined as the ability of the soil to provide, through synergy of physical, chemical and biological factors, conditions for growth and development of plants. It is the main task for the farmer to maintain the fertility of his soil as the basic means of production in agriculture.

In the agriculture sector, the environmental aspect of sustainability is often considered in its bio-physical or agronomical aspects. Thus, this concerns both the impact of the agrosystem on the environment and the system management. Indicators such as nutrients balances, organic matter balance, (bio)diversity, soil cover, soil erosion and compaction affect the soil fertility and functioning of the whole agrosystem. At the same time, they also indicate interaction with the environment and a possible negative impact on it (e.g., nitrogen leaching into ground water, conditions for wildlife, water contamination and other threats caused by eroded soil, etc.). Other indicators, such as energy balance and pesticide use, indicate an interaction with the surrounding environment and resources intensity. However, regarding their names, these indicators may overlap with those listed in the previous part of the chapter; there is an important difference in the organisational level which they describe and thus in their purpose and use.

In part 2 of the chapter indicators for use on the level of country are discussed and also data for calculation of these indicators are collected on state level. Therefore it is not possible to identify farms or even fields which may be the source of possible particular environmental, economic or social problems. Indicators on this level are used as support for political decisions and provide an overall view of development or effect of policy on development of countryside and spread of particular ways of management. In this part of the chapter we deal with indicators calculated with the use of data of a single field within the farm and the result is information for the farmer, usable for an improvement of the farm management. Examples of the indicators are given in part 3.3 of the chapter.

Earlier research of sustainability indicators for agriculture was conducted on the farm level, therefore it was focused mainly on problems of a physical, chemical and biological nature. However, the sustainability concept involves environmental, as well as economic and social pillars and an institutional (governance) dimension (Hřebíček et al. 2012; Moldan et al. 2012; Valtýniová and Křen 2011). Therefore indicators for these issues (which are much more complex regarding type of farm activity) are also used.

The implementation of a comprehensive analysis of farming systems sustainability requires the processing of large amounts of information of a different nature and the use of indicators of different types. These procedures have recently included:

- *Indicator selection and data gathering*. A selection of relevant indicators based on strict quality criteria and accurate data gathering to calculate empirical values of these indicators are an essential element of this kind of study. In order to manage the huge amount of possible indicators and data required, it is recommended that a solid theoretical framework be utilised.
- *Normalization of indicators*. Transforming base indicators into a-dimensional variables (normalization) is required before any aggregation (i.e., to make indicators mathematically operational) is performed. For this purpose, the use of a multiple-attribute utility theory and reference values (sustainability levels that determine the minimum/maximum values of the indicator values) are suggested.
- *Weighing indicators*. Since sustainability is a "social construction", in order to determine the overall sustainability function, it is convenient to take into account society's preferences in order to assign different importance to each dimension/indicator included in the composite indicator. A sensitivity analysis is also advised, with the aim of determining the extent to which weights influence results.
- *Aggregation of indicators*. Although there is a wide variety of functional forms that permits indicators to be aggregated, it is worth taking into account the possible incommensurability of different indicators or dimensions of sustainability.

3.1 Methods for Farming-System Sustainability Assessment

The most common principle of agri-environmental indicators calculation is to compare inputs and outputs of the agrosystem and calculate the balance as it is assumed that all substances which are removed from the system in the harvested biomass have to be returned to the system in appropriate form and amount to prevent exploitation of soil fertility or loses of nutrients or other substances into the environment.

A number of European states are interested in including an Input–Output Accounting (IOA) system as part of EU agro-environmetal support schemes to prove subsidies payments. IOA systems including Green accounts typically use a set of indicators to express the degree of environmental impact of a farm (Goodlass et al. 2003; Halberg et al. 2005; Valtýniová and Křen 2011).

From the indicator methods for farming sustainability assessment we can mention here, for example, the German KUL/USL (Eckert et al. 2000), the KSNL (Breitschuh et al. 2008), REPRO (Hülsbergen 2003) and German Agricultural Society (DLG—Deutsche Landwirtschafts- Gesellschaft) sustainability standards (Schaffner and Hövelmann 2009), the French Indigo method (Girardin et al. 2000) and the Swiss SALCA (Nemecek et al. 2011a, 2011b) and RISE (Grenz et al. 2009) methods.

All these methods are based on the same principle, which utilises indicators. However, each method contains its own set of indicators regarding their number, focus, normalization, weighing and aggregation.

Although all the methods should assess the sustainability of farm performance, most of them do not cover all the three sustainability aspects. With some generalization it can be stated that the formerly developed methods focus on environmental effects (Indigo, KUL/USL, SALCA) or include some economical assessment (REPRO). The social issues have been included since about the year 2000 (KSNL, DLG sustainability standards, RISE).

What also makes important difference, from the users' point of view, is whether the method is primarily a certification procedure (DLG sustainability standards, KSNL) or advisory tool (RISE). In the first case, a farmer just receives the confirmation of the fulfilment of certain criteria (certificate) or also an analysis of weaknesses of his farm management, suggestions for improvement, or he has the possibility of testing these proposals on a computer model of the farm.

3.1.1 Determination of the Threshold Value

Indicators are defined as "observations related to their corresponding reference point" (Riley 2001). The result of measurement or calculation must be interpreted vis à vis the reference value.Van der Werf and Petit (2002) concluded that the identification of *thresholds* which define *the sustainability value* is one of the most critical steps in evaluations carried out with indicators.

The reference value can be a threshold, a norm or a target which is expressed absolutely or relatively. It should be based on research results or relevant statistics and in some cases it should take regionally specific conditions into account.

For example, the reference value for indicators of nitrogen balance is for many users zero as it is assumed that it indicates a stable status of a system. But such implicit value is often criticised due to a lack of scientific arguments confirming dependence between nitrogen surplus and nitrogen loss by leaching which should be indicated by the balance.

However, the reference value is frequently not set up by scientists or specialists but is determined by the stakeholders. In many cases it is useful, when the reference value is the result of an interaction between both groups. Therefore, determining the optimal target value can differ according to the region, and can also be influenced by political impacts.

3.1.2 Normalization of Indicator Values

Some methods report the indicator value in the original units of measurement; others convert the result to a relative figure, which serves as a grade, points or score. As mentioned before, the reference value is necessary for interpretation of indicator values. The normalization process incorporates the reference value of the indicator and the final value already provides an evaluation of the result.

This transformation also enables a further aggregation of indicators also in case they have different original units of measurement. Furthermore, it enables the quick identification of the agrosystem's weak points in a kind of overview of the assessment results. The example is shown in Fig. 1.

However, this method of indicators assessment is more suitable for a complex view, for certification or for policy-makers. If it should be used by farmers as a tool for agrosystem improvement, the original indicators values have to be kept and analysed on this (farm) or even on a lower level (field, crop, etc.), (Bockstaller et al. 2009; Goodlas et al. 2003).

Nardo et al. (2008) have presented a list of normalization methods for the OECD: ranking; Z-score standardization; min–max normalization; distance to a reference measurement; category scale; transformation of indicators above and below the mean; cyclical indicators; balance of opinions; and percentage of annual differences over consecutive years. These algorithms differed for the approach used in the handling of the statistical distribution of original data, with advantages and disadvantages in the normalization process (Freudenberg 2003).

Castoldi and Bechini (2010) presented the basic example of functions (Fig. 2) used to convert bio-physical variables into scores, which represent different qualitative levels of sustainability. Sharp boundaries (e.g., nitrate concentration of groundwater of 50 mg NO_3—-N/l) can be used to divide the indicator output in two ranges, assigning a double judgement: sustainable or non-sustainable (Fig. 2a).

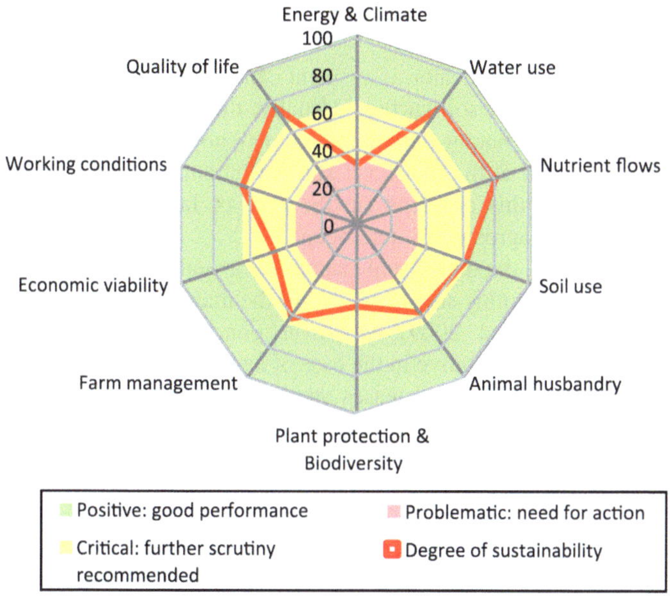

Fig. 1 Overview of results of assessment—method RISE. (Grenz et al. 2009)

Other proposed scores are based on step functions (Fig. 2b) that have an abrupt change at the breakpoints. Alternatively, a gradual variation of sustainability can be approximated by a continuous simple linear function that gradually converts the indicator value into a sustainability score, without abrupt breakpoints (Fig. 2c). One last option is to represent variations of sustainability with a continuous curve (e.g., a bell shape, a logistic curve; Fig. 2d).

The selection of scale, evaluating conversion functions and the range of values are subjective and eventually depend on individual considerations, and therefore need discussion, which is important for effective communication. In any case, the selection should be explicit and transparent (Bockstaller et al. 2009).

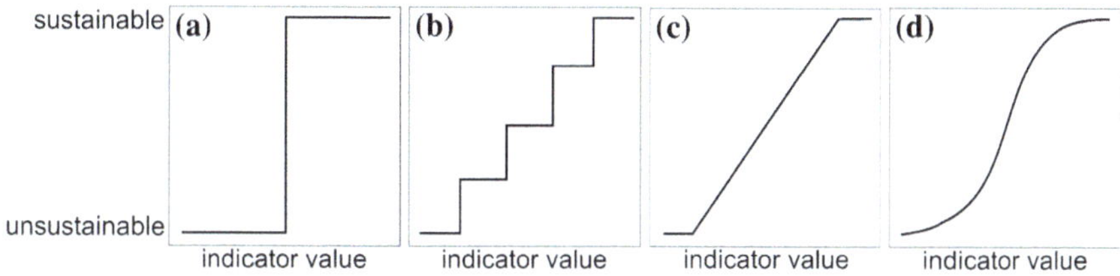

Fig. 2 Different relationships between the value of an indicator and the evaluation of corresponding sustainability: **a** dichotomic judgment; **b** step function; **c** continuous linear function; **d** continuous non-linear function. (Castoldi and Bechini 2010)

3.1.3 Temporal and Spatial Dimension of Indicator Assessment

The spatial boundary for indicator assessment can be set up equal to the boundary of a farm, a plot (field) or soil surface. Usually, the level of input data determination and the level of output data application is not the same. Most often, the input data are collected at the plot level and the outputs are aggregated up to the farm level. However, this can lead (in the case of large farms) to the loss of information about the agrosystem heterogeneity.

With regard to time, the most usual standard is a year; however, methods based on models also use a monthly step (SALCA) or the main stages of the growth cycle (Indigo), (Bockstaller et al. 2009). Moreover, concerning the temporal scale, a single season is not representative in the case of agriculture, due to the inter-annual variability (weather conditions, diseases and pests pressure, variability of prices, etc.). Data from at least 3 years are needed to enable more relevant assessment (Křen et al. 2011; Valtýniová 2011).

3.2 Sustainability Reporting in the Agriculture Sector and Information Technology

3.2.1 Sustainability Reporting

Sustainability reporting is the practice of measuring, disclosing and being accountable for organizational performance towards the goal of sustainable development and is considered synonymous with other terms used to describe—for the purposes of accounting—for economic, environmental, social and governance impacts of farms (Hřebíček et al. 2012). At present, organizations in the agriculture sector have varied approaches to sustainability reporting. While a number of benefits have been proposed for organizations to monitor and report information on performance, there are also several risks and barriers inhibiting open disclosure of facts and figures.

There are a number of barriers and reasons why organizations in the agriculture sector have not embraced sustainability reporting.

Especially in the case of small and medium farms in the Czech Republic, this form of reporting is not supported, or it is markedly heterogeneous (Hřebíček et al. 2009; Chvátalová et al. 2011). We have discovered the main barriers why agriculture organizations do not support sustainability reporting:

- Collecting and managing data is expensive, technical issues with data collection are also a problem.
- Determining a set of appropriate sustainability indicators to monitor and measure is difficult.
- Difficulty in capturing reliable data information (some aspects of the agrosystem are very difficult to collect meaningful and repeatable data).

- Disclosure can create business risks which competitors and regulators may seize upon.
- Difficulty to determine the sphere of influence of an organization.
- Many organizations have good intentions, but simply have not allocated enough resources due to the current economic situation in the Czech Republic.
- Reporting is seen as a superfluous and burdening activity.

The core of these barriers is the certain time-demanding nature of the agriculture farm data-processing, and the absence of positive feedback (Hodinka et al. 2012).

Reporting sustainability indicators does not require the discernment for whom the reporting is destined (obligatory, voluntary), nor does it require the knowledge whether it is designated for organization controlling, public or for public administration. Emphasis, however, is placed on standardizing the descriptive indicators and on the character of the distribution form. In this area, the eXtensible Business Reporting Language (XBRL) standard (XBRL 2012) has taken root as a suitable form of reporting-message distribution (Graning et al. 2011). To a large degree this approach simplifies the exchange and distribution of data and information.

3.2.2 Applied Information Technology for Reporting

We suggest the formalization of the reporting systems of agriculture organizations on the basis of the universal markup language XML by means of the use of the XBRL to minimize above barriers. The XBRL is a worldwide fostered open standard which is supported by the majority of business organizations, financial institutions, investors, regulatory institutions (e.g., SEC, Eurostat, CEBS) and governments. The proposed unified structure of the document complies with the Global Reporting Initiative (GRI) standards (GRI 2006; G3 2006; G3.1 2011). The GRI is a very important network-based organization that produces a comprehensive sustainability reporting framework that is widely used around the world. The GRI has pioneered the development of the world's most widely used sustainability reporting framework in 2000 and is committed to its continuous improvement and application worldwide. The GRI drives sustainability reporting by all organizations.

The XBRL is a widely accepted data standard (XBRL 2012) which enables the exchange of unified financial information between computer systems, software applications and users (Hodinka et al. 2012). It is based on the XML (eXtensible Markup Language) and using it, users will be able to employ a single technology for various applications. The XML language has been put together and standardized by the World Wide Web Consortium (W3C).

The strength of the XBRL springs from its structure, which is divided into an instances document and a taxonomy group. The instances document contains commercial facts which are being reported.

The structure of the XBRL is derived from the Financial Reporting Taxonomies Architecture (FRTA) concept (FRTA 2005) and Financial Reporting Instance Standards (FRIS), (FRIS 2004). It can be characterized by taxonomy schemes and by so-called linkbases. By taxonomy we mean the standardized XML scheme (XSD), which contains concepts of what kind of data will appear in the message. The scheme group describes the syntax, as well as the interconnectedness of each message or its distinct parts. The linkbases are collections of links which enrich the syntax by means of certain semantics (Isenmann and Gomez 2009). The linkup on FRTA principles brings a specification for the appropriate construction and structuring of the sustainable development messages.

For the sustainability reporting concerned with the area of finance, there have been several business use cases. On the basis of the FRTA and FRIS concepts we have built a report dealing not only with the financial side of the organization but also with other aspects, provided documents are correctly edited and amplified. For each document we have created a set of files with the aim of helping the user understand the structure and purpose of the individual cases that correspond with the mentioned standards.

Contents of the template:

- Built taxonomy.
- Display of taxonomies for the presentation, calculation and definition layers.
- Instances document.
- Input data that can be read by the user (Excel file).
- Style file that will define the appearance of the instances document for its conversion into HTML.
- XSL-FO for conversion into PDF and Excel.
- Validation of calculations which can appear in the document.
- Validation of messages.
- Final exporting of the instance document into PDF, HTML and Excel formats using the style files.

By implementing the XML scheme, the agriculture organization gains a whole set of advantages. The administration and editing of information is much easier and much more effective. Employing the abovementioned framework enables an improved communication and collaboration with target groups and concerned parties. By implementing the scheme the company acquires the possibility of creating and publishing compact, focused messages that are generated automatically on the basis of the template rules of one single scheme.

Our approach is based on the GRI taxonomy scheme (G3.1 2011), in which we may define, according to the XBRL specification, new units, which do not appear in financial reporting. The factual values connected with their unit and classified with corresponding contexts, related to the time-period and certain dimension, create the basis for a user-readable and understandable report, which—owing to the style template that has been created—enables the exporting of the message into a text format. This approach will enable the improvement of environmental reports vis á vis their comparability, standardization and reliability, using the latest innovations in the world of financial reporting.

We see another possibility of development in what the XBRL brings for validation and calculation operations in the framework of the report. What is a key factor is a completely pre-set form and structure of the reports or their parts, such as the indicators including the unit specifications and nomenclature. Without such a structure the message-creation process cannot be fully automatized and synchronized with reports published by other parties.

3.3 Overview of Agri-environmental Indicators Used in the Europe on Farm Level

The basic and the most frequently used bio-physical indicators are: *balance of nutrients (nitrogen and phosphorus), organic matter balance* and *energy balance*. The *assessment of pesticides use, agrosystem (bio)diversity* and *soil-protection* are frequently included as well.

Most of the complex methods use relatively simple procedures of indicator calculation in order to reach better feasibility of a method. It appears, generally, that the risk of errors in using a method increases with its complexity. Equally, the demand for input data is increasing (Bockstaller et al. 2009).

The other reason is that, due to a practical feasibility of assessment, input data should only include current agronomical reports and eventually basic characteristics of the locality including information about the soil, the character of terrain relief, climate etc.

The analysis of nutrients management is most frequently oriented on the nitrogen (N), and less frequently to the phosphorus (P), though agriculture can significantly contribute to the eutrophication of water ecosystems. The potassium (K) is mostly ignored. It is not generally a limiting element for water quality but K is important for a long-term soil fertility and production quality (Öborn et al. 2005). Moreover, the interest in optimization of P and K balance is substantiated by the fact that these nutrients originate from limited, non-renewable resources (Bassanino et al. 2011).

The balance is the basis of indicators which deal with nutrients. In all three nutrients (N, P, K), it is based on the same principle of difference between inputs and outputs (Bassanino et al. 2011; OECD 1997). However, in the case of the nitrogen, more possible inputs into the agrosystem can be considered, as well as more ways of its changes and losses compared to the phosphorus and potassium.

The list of basic indicators concerning the nitrogen is shown in Table 7, where examples are given of methods used in the Europe (REPRO, DLG, KUL/USL, KSNL, Indigo, SALCA).

The balance of organic matter is based on the differences between inputs and loss of soil organic matter by mineralization. The level of mineralization depends

Table 7 List of indicators of N balance

Indicator (unit)	Calculation	Input data	Examples of methods in EU
Balance (kg/ha)	Difference of all N inputs and amount of N going away in products (corrected for change of supply)	All inputs, Yields, N content in production	REPRO, DLG, KUL/USL, KSNL
Efficiency of N use (%)	(inputs–outputs)/outputs * 100 %	All inputs, Yields, N content in production	REPRO
Risk of emissions (kg/ha)	Potential model N loss minus effect of measures on loss reduction	Fertilizer inputs Crop Soil type	‘Indigo, SALCA, REPRO

on the grown crop, intensity of soil tillage, soil quality and weather, which are taken into account to a different extent. The established equivalents with defined contents of the carbon and the nitrogen are often used in German methods for the expression of organic matter level (Humuseinheiten—HE (Hülsbergen 2003) or t Reproduktionsfähige organische Substanz—tROS (Eckert et al. 2000). Other frequent equivalents are the dry matter of organic substance or the amount of oxidisable carbon (Cox). The list of indicators of the organic matter balance is presented in Table 8, where examples are introduced of methods used in the EU (KUL/USL, KSNL, REPRO, DLG, Indigo).

The energy assessment is a significant objective indicator of the efficiency of agricultural production (Neudert 1998). The advantage of this approach is that different forms of inputs can be conveyed to the same units (Christen and O’Halloran-Wietholz 2002) and different kinds of production and greatly different ways of production can objectively be compared (Halberg 1999; Refsgaard et al.

Table 8 List of indicators of organic matter balance

Indicator (unit)	Calculation	Input data	Example of methods
Organic matter balance (HE or tROS/ha)	Difference between input of organic matter (in fertilizers and plant residues) and its loss (according to effect of crops)	Organic fertilization, straw management Crop	KUL/USL, KSNL, REPRO, DLG
Supply of organic matter (%)	Loss/inputs of organic matter	Organic fertilization, straw management Soil type	REPRO
	Average dose of organic matter in the last 4 years/recommended dose of organic matter based on content of clay and CaO in soil	Organic fertilization, straw management Soil type, content of clay particles, CaO content	Indigo

1998; Tellarini and Caporali 2000). Different methods can be used for the calculation of plant production energy balance depending on the objective of the analysis performed. The methods mentioned in the literature differ in the spatial and time definition of agrosystem boundaries, in flows of substances and energy, which are taken into account, and in energetic equivalents established for these flows (Kalk and Hülsbergen 1997).

The list of indicators of energy balance is presented in Table 9, where examples are introduced of methods used in the Europe (Indigo, REPRO, KUL/USL, KSNL, DLG, RISE).

The indicator assessing the use of pesticides (see Table 10) is sometimes included into the complex methods but frequently, because of its complexity, builds an independent method. In this case, there is the most expressed variance from simple indicators (of the type of average applied dose of active substance per hectare) to complex models, which also include the persistence period in the environment, the toxicity of substances for particular components of the environment and groups of animals. All indicators for this area use some form of score (Reus et al. 2002). A relatively great number of indicators also includes the component assessing the system of plant protection or non-chemical ways of protection. Indicators also exist which only assess this aspect.

Table 9 List of indicators of energy balance

Indicator (unit)	Calculation	Input data	Example of methods
Consumption of fossil energy (GJ/ha)	Consumption of direct and indirect energy (energy for production of machines and pesticides, eventually fertilizers)	Mechanized work operations and their parameters, inputs of pesticides or fertilizers	Indigo, REPRO
	Consumption of direct energy	Mechanized work operations and their parameters	KUL/USL
Energy balance (GJ/ha)	Difference of energy outputs in products and inputs (consumption) of energy	Mechanized work operations and their parameters, inputs of pesticides or fertilizers Crop yields	REPRO, DLG, KUL/USL, KSNL
Energy efficiency (index)	Ratio of energy outputs in products to inputs (consumption) of energy	Mechanized work operations and their parameters Inputs of pesticides or fertilizers Crop yields	REPRO
Energy need (points)	Combination of energy consumption in factual form (Driving force) and efficiency of its use and independence of a farm with regard to energy sources (State).	Consumption of energy, impacts on the environment Consumption of energy per worker, level of independence with regard to energy	RISE

Table 10 List of indicators assessing the use of pesticides

Indicator (unit)	Calculation	Input data	Example of methods
Index of treatment (index)	Applied dose/approved dose	Pesticide treatment in particular plots including the dose	REPRO
	Treated area/area of farm land in an enterprise		
Intensity of pesticide use(€/ha/year)	Proportion of the pesticides used in an enterprise to the average for a given region	Amount of applied preparations for plant protection	KUL/USL, KSNL
I_{PHY} (points)	Combination of amount and persistence of a pesticide, its penetration to and deposition in the soil and water environment, and air, and its toxicity in the above environments using Fuzzy logic	Date, dose, preparation	Indigo
		Crop	
Plant protection (points)	Combination of crop rotation, amount of active substance and its danger (Driving force) on one side and the system of plant protection and level of management (State) on the other side	Crop rotation, amount of active substances, potential risk of the used substance	RISE
		System of plant protection, education, equipment, existence of waiting periods and buffer zones	
	Amount of active substance * toxicity	Used preparations and their amounts	SALCA
Proportion of untreated area (%)	Proportion of untreated area to the total area of cultivated soil in an enterprise	Untreated area	REPRO
Limitation of risks (points)	Assessment of good agricultural practice (e.g., hand weeding, waste management)	Used non-chemical ways of plant protection	KUL/USL, KSNL

Table 11 List of indicators of system diversity

Indicator (unit)	Calculation	Input data	Example of methods
Diversity of cultures (points)	Indirect assessment through diversity of arable crops and plot size	Crop structure, their distribution in plots, plot size	Indigo
Diversity of crop species (index)	Shannon index	Total area of individual crops in a given year	KUL/USL, KSNL, REPRO, DLG
Biodiversity (points)	Combination of ecological quality of areas and plot size	Zones of ecological compensation, plot size (of low diversity) Area of high diversity	RISE
Mean size of plot (ha)	Median	Acreage of plots	KUL/USL, REPRO
Proportion of ecologically valuable areas (%)	Proportion of agriculturally unemployed ecologically valuable areas to the enterprise cadastre area	Acreage of agriculturally unemployed ecologically valuable areas	KUL/USL, KSNL, REPRO
Proportion of soil long term in rest (%)	Proportion of set aside land to the total cultivated soil area within the enterprise	Acreage of set aside areas	REPRO
Proportion of chemically untreated area (%)	Proportion of untreated area to the total cultivated soil area within the enterprise	Acreage of untreated areas	REPRO

Diversity of an agricultural system (see Table 11) can be considered from several points of view. This can be the diversity of groups or plant species grown in a given year, plot size diversity (Eckert et al. 2000) or the proportion of ecologically valuable area within the farm acreage (Eckert et al. 2000; Grenz et al. 2009). However, the term can also be comprehended differently as the diversity of a farming system concerning the frequency and date of work operations, diversity in soil cultivation, ways of harvest etc. (Schaffner and Hövelmann 2009). Leteinturier et al. (2006) and Thenail et al. (2009) also assess crop rotation, which affects both the stability of the agrosystem, enabling the reduction of inputs of plant protection preparations, and landscape diversity.

Quite often, this area is comprehended from the point of view of the diversity of non-production free living organisms. Actually, it is the original point of view. For example Manhoudt et al. (2005) differentiate biodiversity in crop stand, in field margin stripes, and in stands of line landscape elements.

The information value of indicators assessing spatial and species diversity of the grown crops and the proportion of ecologically valuable areas is influenced by the compactness of the land tenure of the enterprises.

The most frequent field of soil protection assessment is its erosion and compaction. Some authors are also interested in chemical changes characterized by soil reaction changes (Eckert et al. 2000). However, this requires soil analysis; therefore it is indirectly assessed through soil liming (Lewis and Tzilivakis 1998). For the estimate of the soil erosion risk, several procedures have been developed, independently to the sustainability assessment, which are widely used and included in the methodologies for a complex assessment of agricultural enterprises. This is for example the ABAG method (Germany) or USLE (USA). These methods have been adjusted so that they require a relatively large amount of input data, these are nevertheless easily available. The methods assessing the risk of soil compaction require quite detailed information about the mechanization used in each plot (Lebert et al. 2007; Rücknagel et al. 2007). The indicators are listed in Table 12.

4 Example of the Methodology for the Czech Republic: SAGROS

The methodology SAGROS (Sustainable AGROSystems), (Křen et al. 2011, 2012) is the first complex methodology for sustainability assessment for agriculture developed in the Czech Republic. It is based on a long-term research in this field carried out at the Mendel University in Brno since the 90s (Křen and Kostelanský 1996). This research is linked especially to the work of Vereijken (1997), who

Table 12 List of indicators of soil protection level

Indicator (unit)	Calculation	Input data	Example of methods
Potential of erosion (t/ha/year)	ABAG method	Soil type, frequency and intensity of precipitation, length and slope, crop, applied anti-erosion measures	KUL/USL, KSNL
Risk of compaction (index)	Comparison of soil resistance to soil compaction and pressure generated by the machinery used	Soil type Work operation, used machinery	KUL/USL, KSNL
Soil pH (pH)	Direct measurement		KUL/USL, KSNL
Soil management (points)	Combination of intensity of fertilizer and pesticide use and soil load by machinery (Driving force) on one side and soil status (State) on the other side	Soil contamination with fertilizers and pesticides, effect of mechanization Soil state (a) nutrient supply, C content, pH, humidity, salination (b) erosion	RISE

developed a methodology for optimization of farming systems based on cyclic testing and optimization of measures taken within the farm management. The main objective of this approach is valid also for present work. Reaching the sustainable farming practice is based on long-term work with farms, providing them with sufficient information and appropriate advisory services based on scientific foundations. Further development linked to Bockstaller et al. (2009), Hülsbergen (2003), Schaffner and Hövelmann (2009), Valtýniová and Křen (2008, 2011). The SAGROS methodology uses a set of 21 indicators from environmental (8), economic (6) and social (7) dimensions (see Table 13). This set of indicators is similar to those used in the abovementioned methodologies originating in western European countries. The basic framework of the method corresponds with that of Schaffner and Hövelmann (2009) utilized in the DLG sustainability standards. However, calculation was adjusted for the needs and possibilities of the Czech conditions linking to the range of the research done in the country in the issues of individual indicators.

Data sources for the calculations are defined according to records obligatory for the Czech farmers and within the public available statistics and information. This

Table 13 Overview of indicators in SAGROS methodology. (Křen et al. 2011)

Dimension of sustainability	Area of use	Indicator	Unit
Environmental	Water and air	Nitrogen balance	kg/ha
	Resources use	Phosphorus balance	kg/ha
		Potassium balance	kg/ha
		Organic matter balance	t/ha
		Specific energy consumption	MJ/CU
	Diversity	Intensity of plant protection	%
		Soil erosion	–
		System diversity potential	Compilation of 11 indicators
Economic	Stability	Farm income	CZK/ha
		Net margin	CZK/ha
		Indebtedness	%
	Liquidity	Gross margin	CZK/ha
		Liquidity	index
	Profitability	Profitability	CZK/ha
Social	Work and employment	Salary	%
		Working hours	hours/week
		Holidays	Days
		Education and training	%
		Safety and health protection at work	index
		Workers participation	index
	Social engagement	Social engagement	points

CU cereal unit

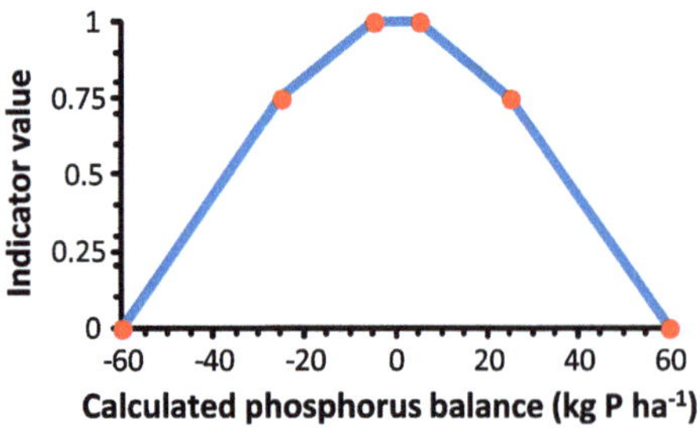

Fig. 3 Example of normalization function for phosphorus balance (Křen et al. 2011)

is important for the practical feasibility of the method and it influences reliability of the results as farmers are not forced to create additional records and analysis only for the purpose of the sustainability assessment. Reference values of the indicators are set according to the legislation and production conditions in the country and in some cases are distinguished regarding the production regions or other conditions. The normalization function is linear, defined by three main points: optimal value (1), the lowest value considered as sustainable (0.75) and unacceptable value (0), (see Fig. 3).

Example of the phosphorus balance calculation:

P_B	P_P_O ($\pm P_C$),
P_B (kg P/ha)	Phosphorus balance for particular field,
P_F (kg P/ha),	Phosphorus input in organic and inorganic fertilizers,
P_O (kg P/ha^{-1})	Phosphorus output (consumption by products harvested from the field),
P_C (kg P/ha)	Correction according to soil analysis (non-compulsory).

The calculation is performed for each field on the farm and these results are aggregated to the farm level using the method of weight arithmetic mean. Information about phosphorus income comes from compulsory agronomic records about fertilization. The value of phosphorus output is calculated based on yield of main- and by-products using standard tabular data about nutrients consumption by crops per unit of the product (Klír et al. 2008). The correction value is based on Work Procedures for Agrochemical Testing of Agricultural Soils in the Czech Republic (Klement 2011). An example of a calculation is given in Table 14.

Table 14 Phosphorus balance calculation for summer barley crop (kg P/ha)

Total P output (P_O)	21.73 kg P/ha
- P consumption - main product (5,3 t/ha)	18.55 kg P/ha
- P consumption - by-product	3.18 kg P/ha
Total P input (P_I)	25.85 kg P/ha
- Organic fertilisers	3.18 kg P/ha
- Straw	3.18 kg P/ha
- Green manures	0 kg P/ha
- Farmyard manures	0 kg P/ha
- Mineral fertilisers	22.67 kg P/ha
P balance (P_B)	4.12 kg P/ha

Calculated value 4,12 kg P/ha corresponds, according to the normalization function, to the optimum range and therefore the indicator final value is 1.

5 Conclusions

The above-discussed and described sets of indicators were created for the implementation of the concept of sustainable development in agriculture, political decisions support and its communication with the public, but also for design and optimization of farms. For this purpose we have selected such indicators that, taken together, provide a comprehensive picture of the examined agrosystem (at the level of user engagement—farm, region, state…) and that individually constitute significant characteristics representing its key features and interaction with the surroundings. The sets of indicators enable in varying degrees to perform a comprehensive analysis of agrosystems focused on sustainability assessment of farming practices at different levels (field, farm, zone, state, region etc.). Individual indicators should provide information about the state of the certain dimension of sustainability (environmental, economic, social and governance).

The sources of data and information used as indicators or for their determination should meet the following requirements:

- high quality statistics, regular monitoring and on a reasonable relationship between costs and predicative ability;
- determination of methodology should be sufficiently exact and reflect the current state of scientific knowledge of the issues described;
- fulfillment of international standards and their usability in modeling or forecasting;
- user-friendliness, which means to be used successfully if parameters of indicators may be logically understood and interpreted;
- contain exact thresholds (reference values) allowing comparison and determination of their evidence.

For the proper interpretation of sustainability assessment results it is necessary to know the methodology of determination of individual indicators and draw conclusions only by respecting the true nature of the indicators used. Despite considerable development in recent years, an objective sustainability assessment still faces some problems. In the systems of created indicators used at a national, regional and EU level, where the character of monitoring prevails, the main problems are the compatibility of indicators and the reliability of used data. For a number of used indicators the absence of the thresholds determination makes the interpretation of the assessment difficult. In addition, at higher levels it is necessary to take into account that the values of the indicators are average figures representing a large land area and involving a large variability and uncertainty of real values. The methods designed for farm level can follow an evaluation of corporate accounting and economic analysis. High quality documentation is already been done in many farms today. Herein farm-wide, comprehensive indicators evaluation is beyond the official requirements but offer a chance for extensive and suitable evaluation of the state of the farm including sustainability assessment.

However, these comprehensive analyses have not yet found a wider introduction in practice. The reasons can be simply summarized in a question "Why in the existing bureaucratic conditions with plenty of legal requirements should further evaluation system be established?" Actually, many working operations on farm must be documented. Usually they are accurate records of the product or the production processes. Gathering the necessary data creates, for the benefit of farm managers a large number of options and supporting documents for practical decision-making. Results of analyses can also be used for the certification of the farm and improvement of its public relations, or to assess the future prospects of the farm.

The possible synergistic effects with other documentation or evaluation systems and the immediate financial benefits should also be emphasized. In the medium term the integration of developed systems of indicators with existing data records appears to be a prerequisite for their adoption by farmers, allowing the possibility of extending their use while minimizing costs. For this purpose attention should be focused on the needs of users, especially farm managers. To realize the target ideas—improving the sustainability of agricultural production—it seems to be a fundamental principle of voluntary participation in the evaluation.

From a broader perspective, for all levels of use of the sustainability indicators (farm, region, state, EU, OECD, FAO) it is desirable to ensure standardization and uniform reporting of the data collection.

All of these are challenges for serious and objective research in this area ensuring the sustainable development of agriculture.

Acknowledgments The chapter was supported by the project No. 2791/2012 "Innovation of the course of Crop Production Systems" and research project No. MSM6215648905 "Biological and technological aspects of sustainability of controlled ecosystems and their adaptation to climate changes", both granted by the Ministry of Education, Youth and Sports of the Czech Republic. Further it is supported by the Czech Science Foundation in the research project No P403/11/2085 "Construction of Methods for Multifactor Assessment of Company Complex Performance in Selected Sectors".

References

Amon B et al. (2011) Analysis of methodologies for calculating greenhouse gas and ammonia emissions and nutrient balances. In: Selenius J, Baudouin L, Kremer AM (eds) Eurostat methodologies and working papers. doi:10.2785/21209

Bassanino M, Sacco D, Zavattaro L, Grignani C (2011) Nutrient balance as a sustainability indicator of different agro-environments in Italy. Ecol Ind 11(2):715–723

Binder CR, Feola G, Steinberger JK (2010) Considering the normative, systemic and procedural dimensions in indicator-based sustainability assessments in agriculture. Environ Impact Assess Rev 30(2):71–81

Bockstaller Ch, Guichard L, Keichinger O, Girardin P, Galan MB, Gaillard G (2009) Comparison of methods to assess the sustainability of agricultural systems. Rev Agron Sustain Dev 29(1):223–235. doi:10.1051/agro:2008058

Breitschuh G, Eckert H, Matthes I, Strümpfel J (2008) Kriteriensystem nachhaltige Landwirtschaft (KSNL). KTBL-Schrift 466. KTBL, Darmstadt

CAP (2012) Legal proposals for the CAP after 2013. Available via DIALOG. http://ec.europa.eu/agriculture/cap-post-2013/legal-proposals/index_en.htm. Accessed 31 Aug 2012

Castoldi N, Bechini L (2010) Integrated sustainability assessment of cropping systems with agro–ecological and economic indicators in northern Italy. Eur J Agron 32(1):59–72

CEC (Commission of European Communities) (1995) Regulation No 79/65/EEC of the Council of 15 June 1965 setting up a network for the collection of accountancy data on the incomes and business operation of agricultural holdings in the European Economic Community. Available via DIALOG. http://eur-lex.europa.eu/LexUriServ/LexUriServ.do?uri=CONSLEG:1965R0079:19951213:EN:PDF. Accessed 31 Aug 2012

CEC (Commission of European Communities) (2006) Communication from the commission to the council and the European Parliament, development of agri-environmental indicators for monitoring the integration of environmental concerns into the common agricultural policy, COM (2006) 508 final. Available via DIALOG. http://eur-lex.europa.eu/LexUriServ/LexUriServ.do?uri=COM:2006:0508:FIN:EN:PDF. Accessed 31 Aug 2012

CEC (Commission of European Communities) (2009) Communication from the Commission to the European Parliament, the Council, the European Economic and Social Committee and the Committee of the Regions, Mainstreaming sustainable development into EU policies: 2009 Review of the European Union Strategy for Sustainable Development, COM(2009) 400 final. Available via DIALOG. http://register.consilium.europa.eu/pdf/en/09/st12/st12453.en09.pdf. Accessed 31 Aug 2012

Charlier H (2010) The EU farm structure surveys from 2010 onwards. http://www.stats.gov.cn/english/icas/papers/P020071115297357509443.pdf. Accessed 31 Aug 2012

Christen O, O'Halloran-Wietholz Z (2002) Indikatoren fur eine nachhaltige Entwicklung der Landwirtschaft. Institut fur Landwirtschaft und Umwelt, Bonn

Chvátalová Z, Kocmanová A; Dočekalová M (2011) Corporate sustainability reporting and measuring corporate performance. In Hřebíček J, Schimak G, Denzer R (eds) Environmental software systems. frameworks of eEnvironment. IFIP advances in information and communication technology, vol 359. Springer, Heidelberg, pp 245–254

DLG (Deutsche Landwirtschafts- Gesellschaft—German Agricultural Society) (2012) Test agriculture. Available via DIALOG. http://www.dlg.org/testsagriculture.html, Accessed 20 Nov 2012

EC (European Commission) (2010) Communication from the Commission. EUROPE 2020. A strategy for smart, sustainable and inclusive growth, COM(2010) 2020 final. Available via DIALOG. http://eur-lex.europa.eu/LexUriServ/LexUriServ.do?uri=COM:2010:2020:FIN:EN:PDF. Accessed 31 Aug 2012

Eckert H, Breitschuh G, Sauerbeck DR (2000) Criteria and standards for sustainable agriculture. J Plant Nutr Soil Sci 163(4):337–351

EEA Glossary (2012) Environmental terminology and discovery service (ETDS). Available via DIALOG. http://glossary.eea.europa.eu/terminology/concept_html?term=sustainability%20indicator. Accessed 31 Aug 2012

EEA Indicators (2012) Indicators and fact sheets about Europe's environment. Available via DIALOG. http://www.eea.europa.eu/data-and-maps/indicators/#c7=all&c5=&c0=10&b_start=0. Accessed 31 Aug 2012

EUROSTAT (2009) Gross Nutrient Balance [aei_pr_gnb]. Available via DIALOG. http://appsso.eurostat.ec.europa.eu/nui/show.do?dataset=aei_pr_gnb&lang=en. Accessed 20 Nov 2012

EUROSTAT (2010) Agri-environmental indicators. Introduction. Available via DIALOG. http://epp.eurostat.ec.europa.eu/portal/page/portal/agri_environmental_indicators/introduction. Accessed 31 Aug 2012

EUROSTAT (2012) NUTS—Nomenclature of territorial units for statistics. Available via DIALOG. http://epp.eurostat.ec.europa.eu/portal/page/portal/nuts_nomenclature/introduction. Accessed 20 Nov 2012

FADN (Farm Accountancy Data Network) (2010) Concept of FADN. Available via DIALOG. http://ec.europa.eu/agriculture/rica/concept_en.cfm. Accessed 31 Aug 2012

FADNCZ (Farm Accountancy Data Network CR) (2011) Farm accountancy data network CR. Liaison agency FADN CZ. Available via DIALOG. http://www.vsbox.cz/fadn/HTM/FADN_V_EU_E.htm. Accessed 31 Aug 2012

FAO (1996) Proceedings of the workshop land quality indicators and their use in sustainable agriculture and rural development. 25–26 Jan 1996, Rome. http://www.fao.org/docrep/W4745E/w4745e00.htm#Contents. Accessed 31 Aug 2012

FAO (2003) Compendium of agricultural—environmental indicators 1989-91 to 2000. Statistics Analysis Service. Statistics Division FAO, Rome. http://www.fao.org/docrep/007/j0945e/j0945e00.htm

FAO (2010) Agri-environment statistics and indicators: current status and future directions. Available via DIALOG. http://www.fao.org/fileadmin/templates/ess/documents/meetings_and_workshops/APCAS23/documents_OCT10/APCAS-10-32_-Agri-envir_01.pdf. Accessed 20 Nov 2012

FAO (2011) Monitoring the interaction between agriculture and the environment: current status and future directions. Available via DIALOG. http://www.fao.org/fileadmin/templates/ess/ess_test_folder/Workshops_Events/AFCAS_22/AFCAS_10b_EN.pdf. Accessed 20 Nov 2012

FAO (2012) Sustainability Assessment of Food and Agriculture systems (SAFA). http://www.fao.org/nr/sustainability/sustainability-assessments-safa/en/. Accessed 20 Nov 2012

FAOSTAT (2012) FAOSTAT web portal. http://faostat3.fao.org/home/index.html#HOME. Accessed 20 Nov 2012

Felice VD, Mancinelli R, Proulx R, Campiglia E (2012) A multivariate analysis for evaluating the environmental and economical aspects of agroecosystem sustainability in central Italy. J Environ Manage 98(5):119–126

Freudenberg M (2003) Composite indicators of country performance: a critical assessment. OECD Sci Technol Ind Working Pap 16: . doi:10.1787/405566708255

FRIS (2004) Financial reporting instance standards 1.0. Available via DIALOG. http://www.xbrl.org/technical/guidance/FRIS-PWD-2004-11-14.htm. Accessed 31 Aug 2012

FRTA (2005) Financial reporting taxonomies architecture 1.0. Available via DIALOG. http://www.xbrl.org/technical/guidance/FRTA-RECOMMENDATION-2005-04-25.htm. Accessed 31 Aug 2012

FSS (Farm Structure Survey) (2012) Farm structure. http://epp.eurostat.ec.europa.eu/statistics_explained/index.php/Farm_structure. Accessed 31 Aug 2012

Girardin P, Bockstaller Ch, Van der Werf H (2000) Assessment of potential impacts of agricultural practices on the environment: the AGRO*ECO method. Environ Impact Assess Rev 20(2):227–239. doi:10.1016/S0195-9255(99)00036-0

Goodlass G, Halberg N, Verschuur G (2003) Input output accounting systems in the European community—an appraisal of their usefulness in raising awareness of environmental problems. Eur J Agron 20(1):17–24. doi:10.1016/S1161-0301(03)00068-6

Gómez-Limón JA, Sanchez-Fernandez G (2010) Empirical evaluation of agricultural sustainability using composite indicators. Ecol Econ 69(5):1062–1075. doi:10.1016/j.ecolecon.2009.11.027

Graning A, Felden C, Piechocki M (2011) Status quo and potential of XBRL for Business and information systems engineering. Bus Inf Syst Eng 3(4):231–239. doi:10.1007/s12599-011-0169-1

Grenz J, Thalmann C, Stämpfli A, Studer C, Häni F (2009) RISE—a method for assessing the sustainability of agricultural productionat farm level. Rural Dev News 1:5–9 http://www.agridea-international.ch/fileadmin/10_International/Petra_ab0508/PDF/RDN/2009_01/1_rise.pdf. Accessed 31 Aug 2012

GRI (2006) GRI reporting framework. Available via DIALOG. http://www.globalreporting.org/ReportingFramework/ReportingFrameworkOverview/. Accessed 31 Aug 2012

GRI (2011) Sustainability reporting guidelines. version 3.1. Food processing sector supplement. Available via DIALOG. https://www.globalreporting.org/reporting/sector-guidance/food-processing/Pages/default.aspx. Accessed 20 Nov 2012

GSCP (2010) Reference tools of the global social compliance programme. Available via DIALOG. http://www.gscpnet.com/working-plan.html. Accessed 20 Nov 2012

G3 (2006) G3 guidelines. Available via DIALOG. https://www.globalreporting.org/reporting/latest-guidelines/g3-guidelines/Pages/default.aspx. Accessed 31 Aug 2012

G3.1 (2011) G3.1 guidelines. Available via DIALOG. https://www.globalreporting.org/resourcelibrary/G3.1-Sustainability-Reporting-Guidelines.pdf. Accessed 20 Nov 2012

Halberg N (1999) Indicators of resource use and environmental impact for use in a decision aid for Danish livestock farmers. Agric Ecosyst Environ 76:17–30

Halberg N, Verschuur G, Goodlas G (2005) Farm level environmental indicators; are they useful? An overview of green accounting systems for European farms. Agric Ecosyst Environ 105:195–212

Hodinka M, Štencl M, Hřebíček J, Trenz O (2012) Current trends of corporate performance reporting tools and methodology design of multifactor measurement of company overall performance. Acta Universitatis Agriculturae et Silviculturae Mendelianae Brunensis. LX(2):85–90

Hřebíček J, Hájek M, Chvátalová Z, Ritschelová I (2009) Current trends in sustainability reporting in the Czech Republic. In: EnviroInfo 2009. Environmental informatics and industrial environmental protection: concepts, methods and tools. 23. International conference on informatics for environmental protection. Shaker Verlag, Aachen, p 233–240

Hřebíček J et al (2011) Integration of economic, environmental, social and corporate governance performance and reporting in enterprises. Acta Universitatis Agriculturae et Silviculturae Mendelianae Brunensis LIX(7):157–166

Hřebíček J (2012) Current trends in corporate performance evaluation and reporting in the Czech Republic. Int J Energy Environ 6(1):39–48

Hřebíček J et al. (2012) Corporate performance indicators for agriculture and food processing sector. Acta Universitatis Agriculturae et Silviculturae Mendelianae Brunensis LX(4):121–132

Hülsbergen KJ (2003) Entwicklung und Anwendung eines Bilanzierungsmodells zur Bewertung der Nachhaltigkeit landwirtschaftlicher Systeme. Berichte aus der Agrarwissenschaft. Shaker Verlag, Aachen

ISEAL alliance (2010) Setting social and environmental standards v5.0 ISEAL code of good practice. http://www.isealalliance.org/sites/default/files/Standard-Setting%20Code%20v5.0_0.pdf. Accessed 31 Aug 2012

Isenmann R, Gomez M (2009) Advanced corporate sustainability reporting—XBRL taxonomy for sustainability reports based on the G3-guidelines of the Global Reporting Initiative. In: Hřebíček J et al (eds) Towards eEnvironment. european conference of the Czech Presidency of the Council of the EU 2009: opportunities of SEIS and SISE: integrating environmental knowledge in Europe. Prague, March 2009. Masaryk University, Brno, p 32

Kalk WD, Hulsbergen KJ (1997) Energiebilanz—Methode und Anwendung als Agrar-Umwelt-indikator. In: Diepenbrock W (ed)Umweltvertragliche Pflanzenproduktion: Indikatoren, Bilanzierungsansatze und ihre Einbindung in Ökobilanzen. Initiativen zum Umweltschutz, Bd. 5 Zeller Verlag, Osnabrück

Klement V (2011) Pracovní postupy pro agrochemické zkoušení zemědělských půd v České republice v období 2011 až 2016 (Working procedures for agrochemical testing of agricultural soils in the Czech Republic in the period 2011–2016).Central Institute for Supervising and Testing in Agriculture, Brno

Klír J, Kunzová E, Čermák P (2008) Rámcová metodika výživy rostlin a hnojení (The framework methodology for plant nutrition and fertilization). Crop Research Institute, Prague

Kocmanová A; Dočekalová, M. (2012) Construction of the economic indicators of performance in relation to environmental, social and corporate governance (ESG) factors. Acta Universitatis Agriculturae et Silviculturae Mendelianae Brunensis LX(4):195–205

Kruijne R, Deneer JW, Lahr J. and Vlaming J (2011) HAIR2010 documentation; calculating risk indicators related to agricultural use of pesticides within the European Union. Wageningen, Alterra, Alterra-report 2113.1, p 202

Křen J, Kostelanský F (1996) Integrované systémy rostlinné produkce (Integrated systems of plant production). In Proceedings of the international conference Ekologické a ekonomické hospodaření na půdě. Brno, 3–4 Sept 1996, pp 85–95

Křen J, Valtýniová S, Marada P et al (2011) Metodika hodnocení trvalé udržitelnosti systémů rostlinné produkce pro podmínky ČR (Methodology of evaluation of crop production systems sustainability for condition of the Czech Republic). Mendel University, Brno

Křen J, Marada P, Valtýniová S et al (2012) Metodika získávání dat pro komplexní analýzy systémů rostlinné produkce (Methodology of data gathering for complex analysis of the field production systems). Mendel University, Brno

Lebert M, Boken H, Glante F (2007) Soil compaction—indicators for the assessment of harmful changes to the soil in the context of the German federal soil protection act. J Environ Manage 82:388–397

Leteinturier B, Heman Jl, De Longueville F, Quintin L, Oger R (2006) Adaptation of a crop sequence indicator based on a land parcel management system. Agric Ecosyst Environ 112:324–334

Lewis KA, Tzilivakis J (1998) Evaluating a technique used to measure environmental performance within agriculture—case studies. Eco-Manage Auditing 5:126–135

Manhoudt AGE, Udo De Haes HA, De Snoo GR (2005) An indicator of plant species richness of semi-natural habitats and crops on arable farms. Agric Ecosyst Environ 109:166–174

Martin J, Henrichs T et al (2012) Environmental indicators report 2012. Ecosystem resilience nd resource efficiency in a green economy in Europe. Available via DIALOG. http://www.eea.europa.eu/publications/environmental-indicator-report-2012. Accessed 31 Aug 2012

Martins C (2008) Farm structure in Czech Republic—2007. Eurostat Statistics in Focus 98. http://epp.eurostat.ec.europa.eu/cache/ITY_OFFPUB/KS-SF-08-086/EN/KS-SF-08-086-EN.PDF. Accessed 31 Aug 2012

Moldan B, Janoušková S, Hák T (2012) How to understand and measure environmental sustainability: Indicators and targets. Ecol Ind 17:4–13

NACE (2011) List of NACE codes. http://ec.europa.eu/competition/mergers/cases/index/nace_all.html. Accessed 31 Aug 2012

Nardo M, Saisana M, Saltelli M, Tarantola S, Hoffman A, Giovannini E (2008) Handbook on constructing composite indicators. OECD, Paris. http://composite-indicators.jrc.ec.europa.eu/Document/Handbook_FINAL.pdf

Nemecek T, Dubois D, Huguenin-Elie O, Gaillard G (2011a) Life cycle assessment of Swiss farming systems: I. Integrated and organic farming. Agricultural Systems. 104(3):217–232. doi:10.1016/j.agsy.2010.10.002

Nemecek T, Dubois D, Huguenin-Elie O, Gaillard G, Schaller B, Chervet Ch (2011b) Life cycle assessment of Swiss farming systems: extensive and intensive productionII. Agric Syst 104(3):233–245. doi:10.1016/j.agsy.2010.07.007

Neudert L (1998) Využití produkčních faktorů a energetické bilance pěstebních technologií u obilnin. (Use of production factors and energy balance practices in cereals). Dissertation, Mendel University

Öborn I, Andrist-Rangel Y, Askegaard M, Grant CA, Watson CA, Edwards AC (2005) Critical aspects of potassium management in agricultural systems. Soil Use Manage 21:102–112

OECD (1997) Environmental indicators for agriculture Volume 1: concepts and frameworks, Paris, France. http://www.oecd.org/tad/env/indicators. Accessed 31 Aug 2012

OECD (2006) Environment at a glance: OECD environmental indicators 2005, Paris, France. http://www.oecd.org/env. Accessed 31 Aug 2012

OECD (2008a) Environmental performance of agriculture in OECD countries since 1990. Available via DIALOG. http://www.oecd.org/document/31/0,3746,en_2649_37401_48509791_1_1_1_37401,00.html. Accessed 31 Aug 2012
OECD (2008b) environmental performance of agriculture in OECD countries since 1990: other agri-environmental indicators. Available via DIALOG. http://www.oecd.org/greengrowth/sustainableagriculture/40673129.pdf. Accessed 31 Aug 2012
OECD (2008c) Environmental performance of agriculture in OECD countries Since 1990 (published 2008). Complete database. Available via DIALOG. http://stats.oecd.org/Index.aspx?datasetcode=ENVPERFINDIC_TAD_2008. Accessed 20 Nov 2012
OECD (2010) OECD Workshop on agri-environmental indicators. Available via DIALOG. http://www.oecd.org/agriculture/sustainableagriculture/oecdworkshoponagri-environmentalindicatorsleysinswitzerland23-26march2010.htm. Accessed 20 Nov 2012
OECD (2012) Agri-environmental indicators. Available via DIALOG. http://www.oecd.org/tad/sustainableagriculture/agri-environmentalindicators.htm. Accessed 31 Aug 2012
Oenema O, Amon B, Van Beek C et al (2011) Farm data needed for agri-environmental reporting. In: Selenius J, Baudouin L, Kremer AM (eds) Eurostat methodologies and working papers. doi: 10.2785/12758
Panell DJ, Schilizzi S (1999) Sustainable agriculture: a matter of ecology, equity, economic efficiency or expedience. J Sustain Agric 13(4):57–66
Pannell DJ, Glenn NA (2000) A framework for the economic evaluation and selection of sustainability indicators in agriculture. Ecol Econ 33(1):135–149
Payraudeau S, van der Werf HMG (2005) Environmental impact assessment for a farming region: a review of methods. Agriculture, ecosystems and amp. Environment 107(1):1–19
Refsgaard K, Halberg N, Kristensen ES (1998) Energy utilization in crop and dairy production in organic and conventional livestock production systems. Agric Syst 57:599–630
Reus J, Leendertse P, Bockstaller C et al (2002) Comparison and evaluation of eight pesticide environmental risk indicators developed in Europe and recommendations for future use. Agric Ecosyst Environ 90:177–187
Riley J (2001) Multidisciplinary indicators of impact and change. Key issues for identification and summary. Agric Ecosyst Environ 87(2):245–259
Rosnoblet J, Girardin P, Weinzaepflen E, Bockstaller C (2006) Analysis of 15 years of agriculture sustainability evaluation methods. In: Bibliotheca fragmenta agronomica—book of proceedings 11(2):707–708
Roth G (2010) Economic, environmental and social sustainability indicators of the australian cotton industry. http://www.cottoncrc.org.au/files/5190792a-3eb1-422c-97db-9dab00edd1e8/3_03_09_Roth.pdf. Accessed 31 Aug 2012
Rücknagel J, Hofmann B, Paul R, Christen O, Hulsbergen KJ (2007) Estimating precompression stress of structured soils on the basis of aggregate density and dry bulk density. Soil Tillage Res 92:213–220
Schaffner A, Hövelmann L (2009) Der DLG-Nachhaltigkeitsstandard „Nachhaltige Landwirtschaft—zukunftsfähig". DBU Schriftenreihe "Initiativen zum Umweltschutz", 1–8. DBU, Berlin. http://www.preagro.de/Veroeff/DLG_Nachhaltigkeitsstandard.pdf. Accessed 31 Aug 2012
Singh RK, Murty HR, Gupta SK, Dikshit AK (2009) An overview of sustainability assessment methodologies. Ecol Ind 9:189–212
Soukopová J, Bakoš E (2010) Assessing the efficiency of municipal expenditures regarding environmental protection. In: Aravossis K, Brebbia CA (eds) Environmental economics and investment assessment III. Book series: WIT transactions on ecology and the environment, vol 131. Pp 107–119. doi: 10.2495/EEIA100101
Stanners D, Bosch P, Dom A, Gabrielsen P, Gee D, Martin J, Rickard L, Weber JL (2007) Frameworks for environmental assessment and indicators at the EEA. In: Hak T, Moldan B, Dahl A (eds) Sustainable indicators: a scientific assessment, scientific committee on problems of the environment, SCOPE 67. Island Press, London

Tellarini V, Caporali F (2000) An input/output methodology to evaluate farms as sustainable agroecosystems: an application of indicators to farms in central Italy. Agric Ecosyst Environ 77:111–123

Thenail C, Joannon A, Capitaine M et al (2009) The contribution of crop-rotation organization in farms to crop-mosaic patterning at local landscape scales. Agric Ecosyst Environ 131:207–219

UNCED (1992) Agenda 21—an action plan for the next century. In: Endorsed at the united nations conference on environment and development, United Nation Press, New York

UN DESA (2007) Indicators of sustainable development: guidelines and methodologies. Available via DIALOG. 3rd edn. http://www.un.org/esa/sustdev/natlinfo/indicators/guidelines.pdf. Accessed 31 Aug 2012

Valtýniová S, Křen J (2008) Selected indicators of sustainability on organic farms in the Czech Republic. Italian J Agronomy. Rivista di Agronomia. 3(3):359–360

Valtýniová S, Křen J (2011) Indicators used for assessment of the ecological dimension of sustainable arable farming – review. Acta Universitatis Agriculturae et Silviculturae Mendelianae Brunensis LIX(3):247–256

Valtýniová S (2011) Porovnání trvalé udržitelnosti konvenčních a ekologických systémů rostlinné produkce (A comparison of the sustainability of conventional and organic crop production systems). Dissertation, Mendel University

Van Beek C, Pietrzaket S, Heesmans H et al (2011) Characterisation of data collection—processing—reporting for agri-environmental policies in Member States of the European Union. In: Selenius J, Baudouin L, Kremer AM (eds) Eurostat methodologies and working papers. doi:10.2785/20525

Van Der Werf HMG, Petit J (2002) Evaluation of the environmental impact of agriculture at the farm level: a comparison and analysis of 12 indicator-based methods. Agric Ecosyst Environ 93:131–145

Velthof G (2011) Agri-environmental indicators: recommendations for priority data collection and data combination. In: Selenius J, Baudouin L, Kremer AM (eds) Eurostat methodologies and working papers. doi:10.2785/20110

Vereijken P (1994) Designing prototypes. Progress report 1 of the research network on Integrated and ecological arable farming systems for EU and associated countries. AB-DLO Wageningen

Vereijken P (1997) A methodical way of protoryping integrated and ecological arable farming systems (I/EAFS) in interaction with pilot farms. Eur J Agron 7:235–250

Vinther FP et al (2011) Data requirements, availability and gaps in agri-environment indicators (AEIs) in Europe. In: Selenius J, Baudouin L, Kremer AM (eds) Eurostat methodologies and working papers. doi:10.2785/20744

Wilson L et al (2011) Direct and indirect data needs linked to the farms for agri-environmental indicators. In: Selenius J, Baudouin L, Kremer AM (eds) Eurostat methodologies and working papers. doi:10.2785/2081

WCED (1987) Our common future. Oxford University Press, Oxford

XBRL (2012) An Introduction to XBRL. http://www.xbrl.org/WhatIsXBRL/. Accessed 31 Aug 2012

Part II
Trends in Eco-production

Crosslinking Eco-innovation in Service and Manufacturing Industries and Knowledge and Operational Industry Orientation

María-del-Val Segarra-Oña, Angel Peiró-Signes, María De Miguel-Molina and Blanca De Miguel-Molina

Abstract Several researchers have studied eco-innovation in order to understand the defining characteristics of companies that consider the environment to be a priority when innovating. However, most studies have focused on manufacturing industries, whereas the service industry economic model is dominant in developed nations. This chapter assesses the similarities and differences among four groups of service and manufacturing industries: Knowledge Manufacturing, Operational Manufacturing, Knowledge Service and Operational Service. Empirical research on a set of Spanish firms shows that the variables affecting the eco-innovative orientations of firms are similar, but with some key differences. The conclusions may help public policymakers plan actions to strengthen environmental proactivity while innovating according to the industry's characteristics.

Keywords Eco-innovation · Service industry · Manufacturing industry

M. Segarra-Oña (✉) · A. Peiró-Signes · M. Miguel-Molina · B. Miguel-Molina
Universitat Politècnica de València, Valencia, Spain
e-mail: maseo@omp.upv.es

A. Peiró-Signes
e-mail: anpeisig@omp.upv.es

M. Miguel-Molina
e-mail: mademi@omp.upv.es

B. Miguel-Molina
e-mail: bdemigu@omp.upv.es

M. Segarra-Oña · A. Peiró-Signes · M. Miguel-Molina · B. Miguel-Molina
INERTE, International Network Research on Tourism and Environment, Valencia, Spain

M. G. Erechtchoukova et al. (eds.), *Sustainability Appraisal: Quantitative Methods and Mathematical Techniques for Environmental Performance Evaluation*, EcoProduction, DOI: 10.1007/978-3-642-32081-1_5,

1 Introduction

Porter and Van der Linde (1995), Esty and Winston (2006) pointed out the relationship between being sustainably concerned ("green") and competitiveness. Since then, studies in this area have attempted to disentangle the underlying factors of superior performance and sustainable relations (Russo and Harrison 2005; Da Silva et al. 2009; Gázquez-Abad et al. 2011).

Understanding why some firms are going beyond legislation and to examine the defining characteristics of firms that consider the environment to be a priority when innovating. Environmental attitude or environmental management are thus crucial variables to be analyzed when we talk about sustainable development-related aspects (Kemp and Pearson 2009; Tietze et al. 2011; Da Silva et al. 2009). However, social pressure (Kuik et al. 2006; Blischwitz et al. 2009; Kalantari and Asadi 2010), public policies (Chappin et al. 2009; Telle and Larsson 2007; Pohoryles 2010) and environmental regulations (Pirani and Secondi 2011; Rivas and Magadán 2010) are also leading knowledge and research in this direction.

The manufacturing industry and its environmental implications have been widely studied, whereas the service sector has received less attention even though the current economy seems to be mainly service-oriented (Montresor and Marzetti 2011). The service-based economy's rapid growth and greater influence, not only in developed countries (Sharma et al. 2007; Lay et al. 2010) but also in emerging economies, is pushing us towards a higher productivity growth trajectory (Joshi 2010).

Eco-innovation is generally understood as any innovation that reduces environmental damage, but its precise definition is still under review. Carrillo-Hermosilla et al. (2010) compiled 16 definitions for eco-innovation, which suggests that it is a multidisciplinary concept that can be studied from different angles: social issues, public policy and regulations, economic benefits and revenues and strategic and managerial concerns (Kemp 2010; Mossalanejad 2011). However, research on this field concerning industry type is still limited.

There is an important shift away from economies that are dependent upon material resources and low-level workforce to economies that utilize knowledge as the key source of competitiveness and innovation. In addition, manufacturing and service taxonomies have pointed to this differentiation, such as that proposed by the OECD based on direct and indirect R&D intensity.

In this study, we address the relevance of analyzing the differences in manufacturing versus service and knowledge-based versus operational-based companies that carry out eco-innovative activities. As a first step, we present the conceptual framework of the study and specify the hypothesis. Then, we introduce the methodology and data set used in the study. The empirical analysis based on the PITEC[1] Spanish database (2009) onto the analysis and interpretation of the results.

[1] Technological Innovation Panel (PITEC) is a statistical tool for the follow-up of activities on technological innovation of the Spanish companies, result of the team work among Nation Statistics Institute (INE), FECYT and COTEC Foundation, with advising of a university researchers group.

We conclude the chapter with some remarks, limitations and further research orientations.

2 Theoretical Approach

Environmental proactivity and innovation affect the competitive positioning of companies (Hitchens et al. 2005; Esty 2006) by transforming existing markets and creating new ones (Beise and Rennings 2005; González-Benito 2010). De Marchi (2011) studied firms' innovative behavior as measured by R&D investment, while Segarra-Oña et al. (2011) highlighted that eco-innovation is positively affected by the size and export orientation of the firm. However, the effects of eco-innovative activities have not been studied from other important angles.

Indeed, characteristics that should be considered include the higher cooperation and more intense relationship that eco-innovative firms establish with suppliers, the reach of the general theory of global chain support to achieve greater competitiveness sharing resources and knowledge and putting the absorptive capacity into value.

Regulations are affecting the rapid development of this field of study (Hellström 2007; Chappin et al. 2009). The key aspects of businesses turning green (Rennings 2000; Gabaldón et al. 2003; Rehfeld et al. 2007; Hu et al. 2010; Del Río et al. 2011) or how previous innovative levels positively affect the environmental orientation of companies (Jaffe and Palmer 1997; Wagner 2008a; De Marchi 2011; Segarra-Oña et al. 2011) have also been considered.

Carrillo-Hermosilla et al. (2010) addressed how eco-innovation influences new business start-ups and contributes to building a more sustainable society, highlighting the importance of collaboration among different stakeholders. At a deeper level, some facilitators of and barriers to eco-innovative behavior or orientation have also been identified, such as the lack of absorptive capacity or highly educated human resources (Chen and Huang 2009), the maturity of the firm (Molero and García 2008; Cainelli et al. 2011) and an industry's technological level (Peiró-Signes et al. 2011). However, the differences between manufacturing and services industries or knowledge-based and operational-based industries have not been addressed until now.

Forsman (2011) made a significant contribution by comparing patterns of innovative behavior between manufacturing and services, indicating as had Sirilli and Evangelista (1998), that there are no significant differences between them. However, they found some differences regarding the size of firms, service industries and types of innovations.

Cainelli et al. (2011) found a negative relation between environmental innovative strategies and employment, turnover and productivity in services firms. This study, which links eco-innovation and service firms, seems to be an exception but, crucially, it did not focus on linking the eco-innovative orientation of firms with their belonging to manufacturing/services industries or to knowledge-/operational-based industries. The present study bridges this gap by determining the similarities and differences between four industry groups, namely Knowledge Manufacturing, Operational Manufacturing, Knowledge Service and Operational Service. Companies' internal decisions and public policy to promote a greener orientation in their innovative activities are then addressed.

2.1 The Urgency to Address a Proper Approach to Eco-innovation

National eco-innovation policies in EU countries are a key part of sustainable development and economic growth strategies (Burciu et al. 2010; Kemp and Oltra 2011; Berger et al. 2001). Nevertheless, few studies of eco-innovation in services or in knowledge-/operational-based industries have been carried out. The need to address different proactive environmental strategies depending on the type of industry of the company is thus necessary (Sharma et al. 2007; Hipp and Grupp 2011).

In every industry, "cleaner production" and "eco-innovation" are popular topics (Schnitzer 1995). However, environmental innovation has been mainly studied from a manufacturing industry perspective (Wagner 2008b; Ziegler and Seijas-Nogareda 2009; Del Río 2010; Peiró-Signes et al. 2011) due to its closeness to environmental technology issues.

The importance of service in manufacturing industries (Lay et al. 2010), the increasing tertiarization of the economy (Peneder et al. 2003) and its impact on the economic growth (Genaro and Melchor 2010) coexist with the movement in developed markets to switch to a knowledge-based economy. Crosslinking the driving forces and distinctive characteristics of eco-innovative activities in manufacturing/service and knowledge/operational industries thus remains an open field for researchers and cannot be treated as though it has the same bases and characteristics. Therefore:

Hypothesis 1 Industries' eco-innovation orientations differ depending on their characteristics.

Hypothesis 1a Manufacturing and service industries' eco-innovation orientations differ.

Hypothesis 1b Knowledge- and operational-based industries differ.

2.2 Environmental Concern as a Distinctive Firms Characteristic

The evolution of firm activities for the sustainable production of goods and services has been thoroughly studied over the past decade (Machiba 2010). Several authors have assessed what factors influence companies to move towards environmentally friendly decisions and activities. In this line of study, González-Benito (2010) identified the main stakeholders' governmental and nongovernmental dimensions that influence attitude towards the environment, showing that environmental awareness among managers, internationalization and industry play a key role. Further, since Nidumolu et al. (2009) pointed out the capabilities needed to improve the stages to lead a company towards the creation of best-practice platforms of change, there is general agreement about environmental management as a social variable (see Sartorius 2006).

Further, Biondi et al. (2002), Berkhout (2005), De Marchi (2011), Segarra-Oña et al. (2011) all studied eco-innovation drivers, concluding that size, export orientation and innovative R&D activities are crucial to eco-innovative development at the firm level. However, few studies have used the reverse perspective to assess what types of firm characteristics establish the eco-innovative orientation and activity of a company. Therefore:

Hypothesis 2 Knowledge- and operational-based industries' eco-innovation orientations differ.

From the second hypothesis, we can expect different characteristics to drive the environmental orientations of firms in different industries. Thus:

Hypothesis 3 Firms that have an environmental orientation differ depending on the type of industry.

3 Research Methods

3.1 Data Collection

Data for this study were collected from the PITEC database, a statistical tool that monitors the technological innovation activities of Spanish companies. The (Spanish National Statistics Institute) INE maintains this database with the advice of academics and experts. The first available dataset is from 2004, and it is updated yearly to include a comprehensive list of Spanish companies that are characterized by the type of innovation they undertake (classified by the Oslo Manual 2005), by industry (in line with the Spanish National Activities Classification, CNAE) and by geographical location. A total of 255 variables are analyzed in the database. Affiliate-level information is not available, as data are taken from an anonymous macroeconomic survey.

We used the last dataset available (2009) to analyze 7,798 firms from a total of 12,817 across all types of industries included in the database. In total, 2021 firms were dropped from the first survey in 2003, although they remain in the database in order to maintain the identification number over time. Therefore, of the 10,796 remaining firms, 2,994 were omitted because they could not be classified as services or manufacturing (e.g. the building/construction sector) or had not answered the main question regarding their environmental orientation when innovating.

We also disregarded those cases that had a lack of data on the variables used in this study.

To evaluate the environmental orientation of the firm while innovating we used the variable Objet11. Objet11 in the PITEC database measures how essential it is for firms to improve their environmental impact while innovating. The PITEC database considers the importance of environmental impact improvement by firms when innovating as particularly important (1) important (2) not so important (3) and not considered or not important (4) Previous works (e.g. Segarra-Oña et al. 2011) have demonstrated that distinguishing between groups 1/2 and 3/4 is challenging, as there is no clear division among groups. Therefore, we created a dummy variable that distinguishes between orientated (1) and low/unoriented (0) firms. We called this variable Objet11 mod.

Then, we classified firms' environmental orientation as oriented or unoriented. We also classified firms by type of industry: *Knowledge Manufacturing, Operational Manufacturing, Knowledge Service and Operational Service.* Knowledge- and operational-based industries were classified by considering whether the labor force in those industries requires a higher education degree. First, we used the two-digit CNAE classification to classify each industry into one of the four mentioned groups (see Table 1). Therefore, we used the variable "ACTIN", which represents the activity classification number (CNAE) for each firm in the survey to match each firm to the corresponding group.

We also wanted to check for the eco-innovation orientation evolution of each of the groups, so we extracted data on Objet11 mod from 2004 to 2008. We did not consider 2003, as this question was not in the survey at that time. Further, the variable designation was changed in 2008, so previous data were taken from the variable Efecto8, which measured exactly the same issue from 2004 to 2007. Figure 1 shows the percentage of firms according to their environmental orientation for each of the groups.

According to this classification, the firms in each group show different characteristics (Table 2). We added median values to limit the impact of outliers on mean values. As expected, R&D expenses and workforce importance were higher in the knowledge-based firms category. We should emphasize that, in general, service firms showed higher sales as well as investment and size values, which point to the fact that the economy has already turned into a service-based

Table 1 Manufacturing versus Service and knowledge-based versus Operational-based industries classification. CNAE codes

	Knowledge	Operational
Manufacturing	20, 21, 22, 23, 24, 25, 26, 27, 28, 29, 30	10, 11, 12, 13, 14, 15, 16, 17, 18, 31, 32
Service	33, 61, 62, 58, 59, 60, 63, 64, 65, 66, 72, 85, 86, 87, 88	35, 36, 37, 38, 39, 45, 46, 47, 49, 50, 51, 52, 53, 55, 56, 68, 69, 70, 71, 73, 74, 75, 77, 78, 79, 80, 81, 82, 90, 91, 92, 93, 95, 96

Fig. 1 Environmental proactivity evolution. Percentage of firms by type of industry

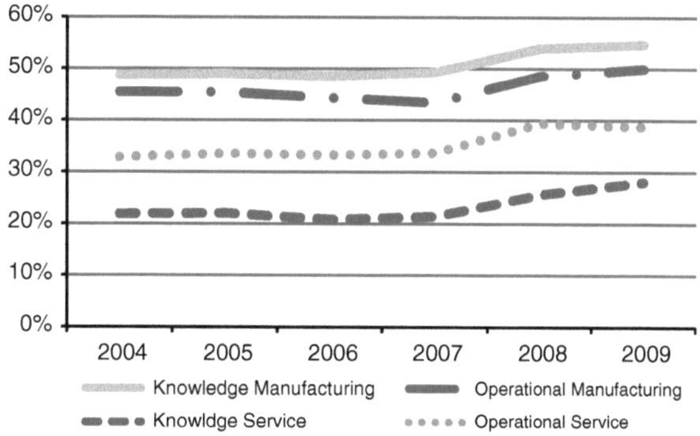

economy. According to theoretical implications, we selected 50 variables (represented in Table 3) to characterize these firms.

As several of these items might show similar constructs, and because underlying relations between these variables have not been specifically identified, we used factor analysis to develop reliable multiple-item measures for each of the underlying theoretical constructs (Hair et al. 1998). Eigenvalues exceeding the accepted cutoff value of 1.0 were not retained in the supplementary data analysis. Together, the 10 retained factors explained about 67.69 % of the variance in the data. In order to increase interpretability we performed a Varimax rotation on the

Table 2 Groups by type of industry

		Knowledge manufacturing	Operational manufacturing	Knowledge service	Operational service
Sales (M€)	Mean	52	37	113	88
	Median	6.9	6.9	3	7.7
	Standard Deviation	265	107	628	467
Investment (M€)	Mean	1.63	1.88	5.45	10.9
	Median	0.06	0.05	0.03	0.02
	Standard Deviation	9	16	42	130
Size (workers)	Mean	166	140	374	583
	Median	44	48	41	80
	Standard Deviation	562	298	1,467	2,495
R&D expenses (M€)	Mean	1.79	0.41	2.5	0.42
	Median	0.153	0.027	0.11	0
	Standard Deviation	9.707	1.72	20.491	2.373
R&D workers	Mean	13	5	17	4
	Median	4	0	2	0
	Standard Deviation	37	12	58	20

Table 3 Selected variables from the PITEC database

PITEC variables	Function type	Explanation
MDOUE	D.	1 = Exports to the EU, 0 = Does not export to the EU
OTROMDO	D.	1 = Exports worldwide, 0 = Does not export worldwide
FINAi (i = 1, 2, 3)	Cat.	Firm receives public financial support from administration at different levels (local, state or national level)
INNFUN	0–100	Basic research
INNAPL	0–100	Applied research
DESTEC	0–100	Technological development
FACEi (i = 1, 2, 3)	Cat.	Importance of external factors in hindering innovation (lack of internal and external financial support, high innovation costs)
FACIi (i = 1,..,4)	Cat.	Importance of external factors in hindering innovation (lack of qualified employees, technologic and market information and cooperation partners)
OTROFACi (i = 1..4)	Cat.	Importance of other factors in hindering innovation (uncertain demand, dominated market, no innovation need)
FUENTEi (i = 1,…,10)	Cat.	Importance of information (universities, papers, …)
INORGNi (i = 1, 2, 3)	Cat.	Organizational innovations (t-2, t)
INCOMNi (i = 1,..,4)	Cat.	Commercial innovations (t-2, t)
OBJET (I = 1,..,10, 14–16)	Cat.	Importance of some objectives (increase market share, increase penetration, cost savings, increase of quality, increase flexibility, increase capacity, energy savings, material savings, …) while innovating

Categorical variables: 1 = High; 2 = Medium; 3 = Low; 4 = Not considered or not important.
Dichotomous variables: 1 = Yes; 2 = No

identified principal components. Then, we assigned items to the factor on which they had the highest loadings. Previously, we eliminated the items that loaded on multiple factors and maintained loadings of at least 0.4 on at least one factor (see Table 4).

We have labeled each of the 10 factors shown in Table 4 for a better understanding of variable grouping.

Since the scales developed via factor analysis are new and have not been validated, we must take care to assess the inter-item reliability of the items comprising each scale. Cronbach's coefficient alpha was thus used to determine inter-item reliability, with alpha values of 0.70 or higher considered to represent acceptable reliability for established scales and 0.60 being acceptable for new scales (Hair et al. 1998). From this analysis, we concluded that the scales comprise reliable items.

Regression factor scores were then extracted to be used in the subsequent analysis. We considered (Hair et al. 1998) cutoff levels, above 0.6 "high" and those below 0.4 "low", to help the factor understanding and labeling rooted in theory. Note that the regression scores have an average of 0, implying that firms show positive evidence that they score over the mean and vice versa.

Finally, in the methodological analysis, we used discriminant analysis to classify the dependent variable, the environmental orientation of the firm, using as predictors a set of independent variables. The discriminant analysis checks whether the selected variables accurately predict the environmental orientations of firms. Therefore, this study establishes a discriminant prediction equation that allows us to classify cases into groups according to their environmental orientation and to assess differences between or among groups. The discriminant function maximizes the differences between the values of the dependent variable in order to differentiate a case into categories of the dependent variable based on the values on the independents.

Structure coefficients show the correlations between a given independent variable and the discriminant scores associated with a given discriminant function. These are used to describe how closely a variable is related to each function. Therefore, four discriminant models were developed to examine if there are differences in the independent variables that describe the environmental orientation in each of the four groups of industries. Models were based on the 10 factors (independent variables) considered and assumed that firms were originally classified into two groups (dependent variable).

4 Analysis and Results

First, we found that manufacturing firms display significantly higher orientations towards environmental aspects while innovating compared with service firms (Fig. 1). As the sample omitted cases that lacked information about the variables needed, we checked the statistics for the whole database, and it followed the same pattern. This result might be due to the strong environmental regulations in manufacturing industries compared with in service industries.

We can also see that Knowledge Manufacturing firms show higher orientations than Operational Manufacturing firms, while on the service side a higher percentage of Operational Services companies have an environmental orientation compared with Knowledge Service companies. By 2009, more than half of manufacturing firms were environmentally oriented, while the percentage for service firms was between 30 and 40 %.

These results show the slow progress towards environmental aspects. Around 6 % of service firms and 5 % of manufacturing firms have changed from not or low oriented to medium or high oriented over a five-year period. However, an average of almost 1 % of companies reverse their orientations towards environmental aspects every year Table 5.

Although Hypothesis 1 shows that manufacturing firms are more concerned about environmental aspects compared with service firms, the slow evolution of

Table 4 Factor analysis results (Varimax rotation)

Components											
Eigenvalues		8.93	4.01	2.60	2.09	1.75	1.42	1.36	1.23	1.20	1.13
% Explained variance		23.49	10.56	6.85	5.50	4.60	3.73	3.58	3.23	3.15	2.98
Factor		1	2	3	4	5	6	7	8	9	10
Internal factors affecting innovation ($\alpha = 0.877$)	FACI2	0.852									
	FACI3	0.837									
	FACI1	0.783									
	FACI4	0.630									
	OTROFAC1	0.621									
	OTROFAC2	0.611									
New product and market development ($\alpha = 0.883$)	OBJET4		0.807								
	OBJET3		0.784								
	OBJET1		0.784								
	OBJET5		0.731								
	OBJET2		0.630								
Cost reduction orientation ($\alpha = 0.885$)	OBJET9			0.819							
	OBJET8			0.803							
	OBJET10			0.799							
	OBJET7			0.702							
	OBJET6			0.652							
Market innovations ($\alpha = 0.763$)	INCOMN3				0.801						
	INCOMN2				0.779						
	INCOMN4				0.702						
	INCOMN1				0.617						
Labor quality improvement orientation ($\alpha = 0.899$)	OBJET14					0.855					
	OBJET15					0.842					
	OBJET16					0.738					

(continued)

Table 4 (continued)

Components						
External information sources ($\alpha = 0.844$)	FUENTE10	0.836				
	FUENTE9	0.799				
	FUENTE11	0.791				
External factors affecting innovation ($\alpha = 0.883$)	FACE1		0.835			
	FACE2		0.831			
	FACE3		0.731			
Organizational innovations ($\alpha = 0.882$)	INORGN2			0.794		
	INORGN1			0.781		
	INORGN3			0.650		
Process innovation ($\alpha = 0.738$)	INNPROC				0.895	
	INNFABRI				0.662	
	INNAPOYO				0.573	
	INNLOGIS				0.454	
Export orientation ($\alpha = 0.776$)	MDOUE					0.864
	OTROPAIS					0.863

Total % of variance explained 67.69. KMO 0.881 sig 0.000

Table 5 Standardized canonical discriminant function coefficients and groups means for service firms

Factors	Knowledge manufacturing	Operational manufacturing	Knowledge services	Operational services
Internal factors affecting innovation	0.052	0.139	0.129	0.077
New product and market development	0.550	0.556	0.183	0.498
Cost reduction orientation	0.744	0.729	0.793	0.674
Market innovations	−0.177	−0.118	−0.140	−0.170
Labor quality improvement orientation	0.506	0.545	0.570	0.602
External information sources	0.412	0.348	0.459	0.452
External factors affecting innovation	0.101	0.056	−0.063	0.079
Organizational innovations	−0.256	−0.208	−0.175	−0.157
Process innovation	−0.192	−0.082	−0.055	−0.060
Export orientation	−0.104	−0.055	−0.147	−0.038
Wilks' lambda	0.666	0.622	0.754	0.663
	$P < 0.001$	$P < 0.001$	$P < 0.001$	$P < 0.001$
Mean scores				
Group 1 (not/low oriented)	0.778	0.779	0.356	0.567
Group 2 (oriented)	−0.644	−0.779	−0.914	−0.896

environmental concern over time is remarkable. Therefore, we explored the characteristics driving the environmental orientations of firms. We first checked the discriminant analysis results for each of the groups. Table 6 shows the coefficients for the discriminant function in each group, as well as the Wilk's lambda and mean scores (Hair et al. 1998). Note that as the predicted variable (Objet11 mod) is a dichotomous variable, there is only one discriminant function in each case.

The coefficient signs indicate the way in which the factor acts. For example, as factor 3 coefficient is positive and factor 3 scores are more negative, the higher the cost reduction orientation is; thus, the discriminant function is more negative, the higher is the cost reduction orientation of the firm. Therefore, we can say that the higher the cost reduction orientation the most likely a firm is to be environmentally oriented. In addition, we present the centroids means for each of the two groups, oriented and low/unoriented, and for each of the four types of industries. The samples had a mean of 0.00 and a standard deviation of 1.00 to allow easy comparisons between groups. The discriminant function scores were standardized. Table 7 shows the relative positions of the cluster means along the discriminant

Table 6 Classification results for the original cases in knowledge manufacturing industries

Knowledge manufacturing		Predicted		
		0	1	Total
Actual	0	**647 (69.87 %)**	279 (30.13 %)	926
	1	240 (21.45 %)	**879 (78.55 %)**	1,119
	Total	887	1,158	2,045
	0	**647 (69.87 %)**	279 (30.13 %)	926
	1	243 (21.72 %)	**876 (78.28 %)**	1,119
	Total	890	1,155	2,045

Maximum chance criterion = 54.72 %; Proportional chance criterion = 50.45 %; Hair et al. Criterion = 63.06 %

axis for each type of industry. The large gap between centroids typifies the highly polarized results shown later.

The structure coefficients (Table 8) are pooled by within-group correlations between the independent variables and standardized canonical discriminant functions in order to allow us to determine the relative importance of each independent variable. The sign of the structure coefficient also shows the direction of the relationship. This statistical technique looks for the statistical significance of functions; however, the discriminant functions should also perform well in classifying firms into their original groups.

Table 9 presents the classification results based on the discriminant function of each group. The existing classification based on the environmental orientation variable value (Objet11 mod) is shown in the rows and the predicted group based on the discriminant functions in the columns. Correct predictions are shown on the main diagonal, while the other cells represent misclassified companies.

These models show a number of similarities. Some variables are crucial in determining the environmental orientations of firms. Cost reduction orientation has the highest discriminant function value in all cases. Labor quality improvement orientation and new product and market development are also crucial in

Table 7 Classification results for the original cases in operational manufacturing industries

Operational manufacturing		Predicted		
		0	1	Total
Actual	0	**905 (74.06 %)**	317 (25.94 %)	1,222
	1	235 (19.23 %)	**987 (80.77 %)**	1,222
	Total	1,140	1,304	2,444
	0	**903 (73.9 %)**	319 (26.1 %)	1,222
	1	238 (19.48 %)	**984 (80.52 %)**	1,222
	Total	1,141	1,303	2,444

Maximum chance criterion = 50 %; Proportional chance criterion = 50 %; Hair et al. Criterion = 62.5 %

Table 8 Classification results for the original cases in knowledge service industries

Knowledge services		Predicted		
		0	1	Total
Actual	0	**778 (73.47 %)**	281 (26.53 %)	1,059
	1	114 (27.6 %)	**299 (72.4 %)**	413
	Total	892	580	1,472
	0	**773 (72.99 %)**	286 (27.01 %)	1,059
	1	115 (27.85 %)	**298 (72.15 %)**	413
	Total	888	584	1,472

Maximum chance criterion = 71.94 %; Proportional chance criterion = 59.63 %; Hair et al. Criterion = 74.54 %

determining the environmental orientation of the firm, but new product and market development is less important for knowledge service firms than for the other groups. Finally, external information arises as the fourth element in importance. We can thus verify Hypothesis 2, as strategic orientation characterizes the environmental orientations of firms to some degree.

The negative coefficients are due, on one hand, to the standardized factor scores and, on the other hand, to the way of categorizing multinomial variables (1 = high to 4 = no relevant). For example, highly environmentally oriented firms have lower scores (negative scores, as regression scores have a mean of 0) than unoriented firms in factors 2 and 3, namely new product and market development and cost reduction orientation while innovating. In other words, highly environmentally oriented firms place more weight on new product and market development and on cost reduction than those firms with less environmental responsibility. By contrast, the discriminant coefficients for market or organizational innovations are negative, while the factor scores for these factors are positive if the firm introduces market or organizational innovations. Therefore, the higher the number of market or organizational innovations the firm has introduced in recent years, the higher is the factor score (positive) and the more negative is the resulting

Table 9 Classification results for the original cases in operational service industries

Operational services		Predicted		
		0	1	Total
Actual	0	**775 (74.88 %)**	260 (25.12 %)	1,035
	1	143 (21.83 %)	**512 (78.17 %)**	655
	Total	918	772	1,690
	0	**770 (74.4 %)**	265 (25.6 %)	1,035
	1	146 (22.29 %)	**509 (77.71 %)**	655
	Total	916	774	1,690

Maximum chance criterion = 61.24 %; Proportional chance criterion = 52.53 %; Hair et al. Criterion = 65.66 %

discriminant function. Since a negative value of the discriminant function reflects the highest chance of being environmentally oriented, we can conclude that firms that introduce market or organizational innovations are more likely to be environmentally oriented that those that do not.

Wilk's lambda tests the significance of each discriminant function. All Wilk's lambda values show significance at a p value of 0.001. In addition, cross-validation classification showed small differences among groups. These results confirm that the model would provide similar results in a wider context.

Note that the hit ratio must be compared to the percentage that would have been correctly classified by chance (50 % if the groups had been equally split) or the expected percentage (50.45 % Knowledge Manufacturing, 50 % Operational manufacturing, 59.63 % Knowledge Service and 52.53 % Operational Service), which is also called the proportional chance criterion. Following Hair et al. (1998) recommendations, we tested classification accuracy by calculating whether the percentage of correct classification was at least 25 % higher than the proportional chance criterion for each discriminant model (1.25 * 50.45 % = 63.06 % Knowledge Manufacturing, 62.5 % Operational manufacturing, 74.54 % Knowledge Service and 65.66 % Operational Service).

As shown in Table 5, the classification accuracies for the estimated models were 74.6, 77.4, 73.2 and 76.2 % for Knowledge Manufacturing, Operational manufacturing, Knowledge Service and Operational Service, respectively, which are higher than the maximum chance criterion and higher than the threshold suggested by Hair et al. (1998). Only Knowledge Service is slightly lower than Hair et al.'s criterion (73.2 Vs. 74.54 %), but we should take into account that the higher the group size difference (higher proportional chance criterion) the harder it is to meet Hair et al.'s criterion. Then, we can consider this to be an acceptable model in this case.

Cross-validation was used over a hold-out sample to further validate the discriminant models as suggested by leading statisticians (Hair et al. 1998). Each case was classified using a discriminant function based on all cases except the given case. As Table 7 shows, these cross-validated cases were classified as exceeding the proportional chance criterion, maximum chance criterion and Hair et al. (1998) criterion, except for Knowledge Service in relation to Hair et al.'s criterion as mentioned before.

These results support the finding of no large differences between manufacturing and service firm characteristics according to their environmental orientations, thereby rejecting Hypothesis 3. Although there was a small difference in the new product and market development factor in knowledge services, the main factors affecting the environmental orientations of firms remain the same regardless of whether they belong to manufacturing or services industries or to knowledge- or operational-based industries. It is remarkable, as we said before, that the models are quite proficient in classifying cases, which indicates that these variables would be particularly useful to discriminate between oriented and low/unoriented firms. This could determine those aspects to be worked out to steer the companies towards environmental concerns.

5 Discussion, Conclusions, Limitations, and Further Research

The increasing importance that society, business and government place of environmental concern makes it crucial to understand the distinctive characteristics of eco-innovations in order to develop specific eco-innovative promotion programs (Peiró-Signes et al. 2011). The objective of this study was to determine, empirically, the similarities and differences between the characteristics of manufacturing and service and knowledge- and operational-based firms according to their environmental orientations. Although the economies in developed countries are increasingly based on services and knowledge, eco-innovation has not been studied regarding these considerations. It is thus necessary to deepen our knowledge of the impact of firm characteristics on environmental orientation to determine the policies and strategies to promote this direction.

The results provide several compelling insights. The first result suggests that manufacturing firms show higher orientations towards the environment compared with service firms, supporting theories that state that the manufacturing industry is leading the green revolution (Jovane et al. 2008; Wagner 2007; Johnstone and Labonne 2009). This could be explained by the regulation applied to manufacturing firms compared with service firms (Demirel and Kesidou 2011; Heyes and Kapur 2011; Mickwitz et al. 2008). In addition, stakeholders' stronger feelings about the environmental impact of manufacturing industries press manufacturing firm managers to focus on environmental aspects more than service managers might (González-Benito. 2010). We also found that knowledge-based manufacturing firms have higher concern about environmental aspects than operational-based manufacturing firms, whereas in services it is the other way round.

Furthermore, the second result of this research confirms previous findings (Segarra-Oña et al. 2011), showing highly polarized positions in environmental aspects. We are able to describe accurately a firm's environmental orientation. An environmental company is highly concerned about cost reduction and labor quality as well as about developing new products and markets and it sees external information sources as relevant. In other words, firms are concerned about internal and external operational improvement, in the same direction that previous studies indicated (Zhu et al. 2006; Dekker et al. 2012), but with empirical demonstration in our case.

Finally, the analysis confirms that knowledge manufacturing, operational manufacturing, knowledge service and operational service firms classified according to their environmental orientations differ little in terms of their operational objectives and priorities. This finding suggests that only minor differences can be explained by the particularities of each category. Further, the identification of the aspects that characterize environmental orientation through the discriminant analysis is encouraging; the presented models classified about 75 % of firms correctly, although the classification task is inherently difficult due to the heterogeneity of the four groups.

Based on these results, it becomes possible to determine which firm behavior has to be promoted in order to focus on environmental aspects. Moreover, similar policies can be considered among the four groups of industries studied. According to the results, environmentally oriented firms are characterized by a dynamic and "open to change" behavior, showing some of the characteristics that can be found in those companies looking for excellence. The results also indicate that service firms' managers should pay attention to the characteristics of manufacturing firms and their actions to improve their environmental orientations, as such firms are ahead of service companies in environmental aspects.

The use of the PITEC database is particularly useful, but eco-innovation is not studied directly. The dependent variable used (Objet11) is defined as the intention to take into consideration the environment when innovating; however, it has not been designed specifically to evaluate eco-innovation. Considering both the implications of the results for a better understanding of green innovation development and the limitations found, future studies should aim to extend the analysis to specific objectives, such as organizational, process, marketing or product eco-innovations among firms as well as to cover specific industries.

Acknowledgments The authors would like to thank the Spanish Economy and Competitiveness Ministry for its support through the research project (EC02011-27369), and the Uni-versitat Politècnica de València for its research funding to the project (PAID-06-2011-1879), "Innovative practices' on environmental firms "performance". Identifying moderating effects". Also the UPV Research Micro-cluster "Globalization, tourism and heritage: towards a sustainable development".

References

Beise M, Rennings K (2005) Lead markets and regulation: a framework for analyzing the international diffusion of environmental innovations. Ecol Econ 52(1):5–17

Berger G, Flynn A, Hines F, Johns R (2001) Ecological modernization as a basis for environmental policy: current environmental discourse and policy and the implications on environmental supply chain management. Innov: Eur J Soc Sci Res 14(1):55–72

Berkhout F (2005) Eco-innovation: reflections on an evolving research agenda. J Int J Technol Policy Manage 11(3–4):191–197

Biondi V, Iraldo F, Meredith S (2002) Achieving sustainability through environmental innovation: the role of SMEs. Int J Technol Manage 24(5–6):612–626

Blischwitz R, Giljun S, Kuhndt M, Scmidt B (2009) Eco-innovation: putting the EU on the path to a resource and energy efficient economy. Wuppertal Institute, Brussels

Burciu A, Bostan I, Condrea P, Grosu V (2010) Financing the environmental policies in the communitarian space. Environ Eng Manage J 9(9):1179–1185

Cainelli G, Mazzanti M, Zoboli R (2011) Environmentally oriented innovative strategies and firm performance in services micro-evidence from Italy. Int Rev Appl Econ 25(1):61–85

Carrillo-Hermosilla J, Del Río P, Könnölä T (2010) Diversity of eco-innovations: reflections from selected case studies. J Cleaner Prod 18(10–11):1073–1083

Chappin MMH, Vermeulen WJV, Meeus MTH, Hekkert MP (2009) Enhancing our understanding of the role of environmental policy in environmental innovation: adoption explained by

the accumulation of policy instruments and agent-based factors. Environ Sci Policy 12(7):934–947
Chen C, Huang J (2009) Strategic human resource practices and innovation performance the mediating role of knowledge management capacity. J Bus Res 62(1):104–114
Da Silva EM, Jabbour CJC, Santos FCA (2009) Integrating environmental management and manufacturing strategy: an emerging competitive priority. Int J Environ Technol Manage 10(3):397–411
De Marchi V (2011) Environmental innovation and R&D cooperation: empirical evidence from Spanish manufacturing firms. Res Policy 41(3):614–623
Dekker R, Bloemhof J, Mallidis I (2012) Operations research for green logistics–an overview of aspects, issues, contributions and challenges. Eur J Oper Res. doi:10.1016/j.ejor.2011.11.010
Del Río P, Carrillo-Hermosilla J, Könnölä T (2010) Policy strategies to promote eco-innovation. J Ind Ecol 14(4):541–557
Del Río P, Tarancón M, Callejas F (2011) Analyzing the determinants of environmental technology investments, a panel-data study of Spanish industrial sectors. J Cleaner Prod 19(11):1170–1179
Demirel P, Kesidou E (2011) Stimulating different types of eco-innovation in the UK: government policies and firm motivations. Ecol Econ 70(8):1546–1557
Esty DC, Winston AS (2006) Green to gold, how smart companies use environmental strategy to innovate, create value, and build competitive advantage. Wiley, Hokoben
Forsman H (2011) Innovation capacity and innovation development in small enterprises a comparison between the manufacturing and service sectors. Res Policy 40(5):739–750
Gabaldón S, López JB, Carda JB (2003) Legislación y gestión medioambiental en la producción de baldosas cerámicas, Boletín de la Sociedad Española de Cerámica y vidrio, (In Spanish) 42(3):169–179
Gázquez-Abad JC, Jiménez-Guerrero JF, Mondéjar-Jiménez JA, Cordente-Rodríguez M (2011) How companies integrate environmental issues into their marketing strategies. I J Environ Res 10(12):1809–1820
Genaro D, Melchor E (2010) The impact of the tertiarization process in Spanish economic growth from a regional perspective. Serv Ind J 30(3):359–374
González-Benito J (2010) Supply strategy and business performance: an analysis based on the relative importance assigned to generic competitive objectives. Int J Oper Prod Manage 30(8):774–797
Hair JF, Anderson RE, Tatham RL, Black WC (1998) Multivariate date analysis, 5th edn. Prentice-Hall, Englewood Cliffs
Hellström T (2007) Dimensions of environmentally sustainable innovation: the structure of eco-innovation concepts. Sustain Dev 15(3):148–159
Heyes A, Kapur S (2011) Regulatory attitudes and environmental innovation in a model combining internal and external R&D. J Env Econ Manage 61(3):327–340
Hipp H, Grupp H (2011) Innovation in the service sector: the demand for service-specific innovation measurement concepts and typologies. Res Policy 34(4):517–535
Hitchens D, Thankappan S, Trainor M, Clausen J, De Marchi V (2005) Environmental performance, competitiveness and management of small businesses in Europe. Tijdschrift-vooreconomische en socialegeografie 96(5):541–557
Hu H, Parsa HG, Self J (2010) The dynamics of green restaurant patronage. Cornell Hospitality Q August 51(3):344–362
Jaffe AB, Palmer K (1997) Environmental regulation and innovation: a panel data survey. Rev Econ Stat 79(4):610–619
Johnstone N, Labonne J (2009) Why do manufacturing facilities introduce environmental management systems? Improving and/or signaling performance. Ecol Econ 68(3):719–730
Joshi S (2010) From the 'Hindu rate of growth' to 'unstoppable India': has the services sector played a role? Serv Ind J 30(8):1299–1312

Jovane J, Yoshikawa H, Alting L, Boër CR, Westkamper E, Williams D, Tseng M, Seliger G, Paci AM (2008) The incoming global technological and industrial revolution towards competitive sustainable manufacturing. CIRP Annals Manufact Technol 57(2):641–659

Kalantari K, Asadi A (2010) Designing a structural model for explaining environmental attitude and behavior of urban residents (case of Tehran). Int J Environ Res 4(2):309–320

Kemp R (2010) Eco-innovation: definition, measurement and open research issues. Economia Politica 27(3):397–420

Kemp R, Oltra V (2011) Research insights and challenges on eco-innovation. Dyn Ind Innov 18(3):249–253

Kemp R, Pearson P (eds) (2009) Final report of the project measuring eco-innovation (MEI). United Nations University (Working paper series), Maastricht

Kuik O, Berkhout F, Venn A, Ekins P, Anderson J, Bassi S, Stantcheva E, Brink P, Ros J (2006) Innovation dynamics induced by environmental policy. Institute for Environmental studies, Amsterdam

Lay G, Copani G, Jager A, Biege S (2010) The relevance of service in European manufacturing industries. J Serv Manage 21(5):715–726

Machiba T (2010) Eco-innovation for enabling resource efficiency and green growth: development of an analytical framework and preliminary analysis of industry and policy practices. In Springer Books Int Econ Resour Effi 5:371–394

Manual O (2005) Guidelines for collecting and interpreting innovation data, 3rd edn. OECD and Eurostat organization for economic co-operation and development statistical office of the European Communities

Mickwitz P, Hyvättinen H, Kivimaa P (2008) The role of policy instruments in the innovation and diffusion of environmentally friendlier technologies: popular claims versus case study experiences. J Cleaner Prod 16(1):162–170

Molero J, García A (2008) The innovative activity of foreign subsidiaries in the Spanish innovation system: an evaluation of their impact from a sectorial taxonomy approach. Technovation 28(11):739–757

Montresor S, Marzetti GV (2011) The deindustrialisation/tertiarisation hypothesis reconsidered: a subsystem application to the OECD7. Camb J Econ 35(2):401–421

Mossalanejad A (2011) The role of economic policy and environment in sustainable development. Int J Environ Res 5(2):395–402

Nidumolu R, Prahalad C, Rangaswami M (2009) Why sustainability is now the key driver of innovation. Harvard Bus Rev 87(9):57–64

Peiró-Signes A, Segarra-Oña M, Miret-Pastor L, Verma R (2011) Eco-innovation attitude and industry's tecnological level an important key for promoting efficient vertical policies. Environ Eng Manage J 10(12):1893–1901

Peneder M, Kaniovski S, Dachs B (2003) What follows tertiarisation? Structural change and the role of knowledge-based service. Serv Ind J 23(2):47–66

Pirani E, Secondi L (2011) Eco-friendly attitudes: what European citizens say and what they do. Int J Environ Res 5(1):67–84

Pohoryles RJ (2010) Europe in the making—what role environmentalism; and, why should sustainable development be less than environmentalism? A réplique to Warleigh-Lack. Innov: Eur J Soc Sci Res 23(4):319–322

Porter ME, Van der Linde C (1995) Toward a new conception of the environment competitiveness relationship. J Econ Perspect 9(4):97–118

Rehfeld K, Rennings K, Ziegler A (2007) Integrated product policy and environmental product innovations: an empirical analysis. Ecol Econ 61(1):91–100

Rennings K (2000) Redefining innovation-eco-innovation research and the contribution from ecological economics. Ecol Econ 32(2):319–332

Rivas J, Magadán M (2010) Less green taxes and more control over pollutant industries: a theoretical proposal. Environ Eng Manage J 9(9):1173–1177

Russo MV, Harrison NS (2005) Organizational design and environmental performance: clues from the electronics industry. Acad Manag J 48(4):582–593

Sartorius C (2006) Second-order sustainability-conditions for the development of sustainable innovations in a dynamic environment. Ecol Econ 58(2):268–286
Schnitzer H (1995) Environment and innovation: introducing cleaner production. Innov: Eur J Soc Sci Res 8(3):309–317
Segarra-Oña M, Peiró-Signes A, Albors-Garrigós J, Miret-Pastor L (2011) Impact of innovative practices in environmentally focused firms: moderating factors. Int J Environ Res 5(2):425–434
Sharma S, Aragón-Correa JA, Rueda-Manzanares A (2007) The contingent influence of organizational capabilities on proactive environmental strategy in the service sector: an analysis of North American and European ski resorts. Canadian J Adm Sci 24(4):268–283
Sirilli G, Evangelista R (1998) Technological innovation in services and manufacturing: results from Italian surveys. Res Policy 27(9):881–899
Telle K, Larsson J (2007) Do environmental regulations hamper productivity growth? How accounting for improvements of plants environmental performance can change the conclusion. Ecol Econ 61(2–3):438–445
Tietze F, Schiederig T, Herstatt C (2011) What is green innovation?—A quantitative literature review. The XXII ISPIM conference. Available at SSRN: http://ssrn.com/abstract=1846882
Wagner M (2007) On the relationship between environmental management, environmental innovation and patenting: evidence from German manufacturing firms. Res Policy 36(10):1587–1602
Wagner M (2008a) Empirical influence of environmental management on innovation: evidence from Europe. Ecol Econ 66(2–3):392–402
Wagner M (2008b) On the relationship between environmental management, environmental innovation and patenting: evidence from German manufacturing firms. Res Policy 36(10):1587–1602
Zhu Q, Sarkis J, Lai K (2006) Green supply chain management: pressures, practices and performance within the Chinese automobile industry. J Cleaner Prod 15(11–12):1041–1052
Ziegler A, Seijas-Nogareda J (2009) Environmental management systems and technological environmental innovations: exploring the causal relationship. Res Policy 38(5):885–893

Trends in ESG Practices: Differences and Similarities Across Major Developed Markets

Angel Peiró-Signes and María-del-Val Segarra-Oña

Abstract Over the past decade, there has been a significant growth in corporate reporting of Environmental, Social, and Corporate Governance (ESG) factors. Stakeholders strategy can be affected by ESG factors, but also by their relationship with economic performance. Companies and government should have a deeper knowledge of the ESG information to be able to manage long-term value creation and to mitigate risk, so opportunities and threats can be identified properly. ESG exposure can be important at a company and at an industry level, but also at a country or area level. In this study we try to identify differences in EGS factors across countries, continents, and industries, using country-level, area-level, and industry-level data. We determine the trends in Environmental, Social, and Corporate Governance factors to conclude which are key to improving these ratings.

Keywords ESG ratings

1 Introduction

The increasing importance of Environmental, Social, and Corporate Governance (ESG) ratings in the investment decision-making process (Kemp et al. 2005) is a reality. Governments, institutional shareholders, and asset managers, among

A. Peiró-Signes (✉) · María-del-Val Segarra-Oña
Universitat Politècnica de València, Valencia, Spain
e-mail: anpeisig@omp.upv.es

María-del-Val Segarra-Oña
e-mail: maseo@omp.upv.es

A. Peiró-Signes · María-del-Val Segarra-Oña
INERTE International Network Research on Tourism and Environment, Valencia, Spain

M. G. Erechtchoukova et al. (eds.), *Sustainability Appraisal: Quantitative Methods and Mathematical Techniques for Environmental Performance Evaluation*, EcoProduction, DOI: 10.1007/978-3-642-32081-1_6,

others, are focusing their interests on EGS ratings, which encompass a wide agenda of issues (Besley and Maitreesh 2007; Vargas-Vargas et al. 2010; De-Miguel-Molina et al. 2011). ESG concept multi-dimensionality makes it more difficult to reach clear conclusions for all these groups of interest.

Previous studies have focused on determining the relationship between EGS scores and economic performance. Some authors suggested that "good" companies with high ESG scores earn positive abnormal returns (Statman and Glushkov 2009) due either to investors underestimating the benefits of ESG or overestimating its costs, mispricing the value relevance of ESG concerns, or compensation for risk. On the contrary, other studies show that some good companies earn negative abnormal returns, also explained as either mispricing or compensation for risk. At the same time, companies in the alcohol, gambling, tobacco, firearms, military, and nuclear industries also earn positive abnormal returns (Statman and Glushkov 2009; Hong and Kacperczyk 2009).

Governments are struggling to determine which kind of policy will increase the economic results and competitiveness of their companies (Duran et al. 2009; Kranjac et al. 2012). In this context, ESG information cannot be ignored by companies and governments if they want to guarantee a correct strategy to face future scenarios (Melnyk et al. 2003).

Additionally, key performance indicators (KPIs) are of a qualitative nature and therefore difficult to express in a comprehensive way in order to evaluate them. Also, these KPIs may vary depending on the rating firm. Bassen and Kovacs (2008) analyzed an instrument to facilitate the quantification and representation of such data, aiming to promote standardized qualitative reporting for extra-financial information.

In this study, we try to give an overview of actual ESG aspects and the evolution in the past five years in order to uncover the trends of ESG company strategies. We also provide suggestions to policy makers in terms of how to face future uncertainty.

2 EGS Ratings

This study is based on Thomson Reuters ASSET4's integrated ratings for the period 2006–2010, a leading provider of environmental, social, and corporate governance (ESG) data.

ASSET4 gathers quantitative and qualitative ESG data on 3000+ global companies and scores them on three aspects: Environmental, Social, and Corporate Governance. We analyze differences and similarities across major markets in ASSET4 data, which includes information about the company and industry. Additionally, we have tracked industry scores to identify which industries are performing better in these scores. In this study, we use the company-level data to analyze differences and similarities across major developed markets and across industries. Although ASSET4 ESG data dates back to 2002, we dismissed data from 2002 to 2005 because of its scarcity.

Normally, reporting is voluntary, so ESG data are made publicly available by the companies themselves. Large companies have the resources to issue corporate social responsibility reports, so these are usually included in the ESG ratings. Assuming that we are narrowing the scope to address more effectively ESG trends (Chatterji et al. 2009), we think the results of the reported companies are likely to represent industry and country trends, as they are facing the same regulations. However, it can be expected poorer ESG performance for small and medium enterprises (SMEs) as they do not have as much resources to measure and report these practices as big companies or corporations.

Companies with the most to hide are the least likely to volunteer poor ESG practices. Therefore, we can expect higher information for industries that do not have many controversies, like service industries, and companies in "exposed" industries that want to "wash" their business in front of stakeholders (Melnyk et al. 2003; Barba et al. 2008; Delmas and Vered 2010).

Furthermore, lower ratings for Environmental, Social, and Corporate Governance measures imply lower disclosure and/or lower adherence to ESG standards, which, eventually, may reflect a riskier and more unstable environment for investors. However, higher economic growth in certain areas and countries, like Asia/Pacific and BRICS, may enable companies to invest more in ESG practices, to limit ESG-related costs' impact in the long term. Then, we should expect a higher ratings increase in those countries in the next years.

Table 1 shows Asset4 universe distribution by countries and industries taking into account the last data available. Countries with less than 30 companies where not included as a single entry on the table. We only reported top industries selected by the number of companies reported in the study.

As we can see, developed countries have a large number of companies tracked by the ESG scores. Nevertheless, new developing countries, like those known as BRICS, are quickly increasing their presence in these scores. There were five BRICS companies included in Asset4 database in 2006, a number which rose to 212 companies in 2010.

On the other hand, companies in industries with a big share in major economies' GPD report more than those industries with a lower share.

Table 2 shows ESG results across five major areas of developed markets. We use ASSET4's company ratings across the Environmental (En), Social (So), and Corporate Governance (CG) to analyze which countries and areas have been the best and worst performers in ESG metrics, based on the latest data available. Additionally, we reported Economic and Equally Weighted scores as they were available for the latest year reported. Economic performance score is based on client loyalty, economic performance, and shareholders' loyalty, and the Equally Weighted scores is an integrated rating which equally weights the scores across the four ratings.

Table 1 Asset4 universe (last data available)

Country	Ner	%	Industry	Ner	%
United States	1,127	28.9	Banking services	291	7.8
Japan	430	11.0	Metal/Mining	248	6.7
United Kingdom	376	9.7	Oil/Gas	200	5.4
Australia	295	7.6	Machinery/Equipment/Components	177	4.8
Canada	291	7.5	Insurance	151	4.1
Hong Kong	116	3.0	Telecommunications services	119	3.2
France	98	2.5	Investment services	116	3.1
Germany	84	2.2	Electric utilities	116	3.1
Switzerland	67	1.7	Software/IT services	115	3.1
South Korea	63	1.6	Chemicals	111	3.0
China	60	1.5	Food/Tobacco	110	3.0
Italy	54	1.4	Commercial services/Supplies	108	2.9
Spain	54	1.4	Real estate operations	106	2.9
Sweden	54	1.4	Media/Publishing	105	2.8
Singapore	52	1.3	Retailers—Specialty	97	2.6
Taiwan	51	1.3	Hotels/Entertainment services	89	2.4
Brazil	48	1.2	REIT—Residential/Commercial	87	2.3
South Africa	46	1.2	Construction/Engineering/Materials als	86	2.3
India	44	1.1	Biotechnology/Pharmaceuticals	82	2.2
Netherlands	41	1.1	Energy related equipment/Services	82	2.2
Malaysia	38	1.0	Automobiles/Auto parts	79	2.1
Russian Federation	33	0.8	Semiconductors/Semiconductor equipment	71	1.9
Others	373	9.6	Others	973	26.2

The percentages in Table 2 represent the degree to which each country, by the aggregation of the company information, has implemented policies and initiatives for Environmental, Social, and Corporate Governance issues.

Table 3 examines ESG scores across industries. The percentages in Table 3 rep-resent the degree to which each industry, by the aggregation of the company information, has implemented policies and initiatives for EGS issues.

We report, in addition to the mean of the mentioned scores, the median values. Median values are not influenced by outliers and might represent better the real position of countries and industries than median values, so we focus on median values to interpret the data.

We highlight in bold letters the best-in-class country or industry for each of the five scores and also the worst-in-class country or industry. In Table 2, we aggregate the countries into four areas or interest groups: Asia/Pacific, Europe, BRICS, and North America. Aggregated results for each area are shown in Table 2 in italics. As before, we highlight Best-in-Class and Worst-in-Class for the four areas.

Then, we use the country-level data and area-level data to analyze differences and similarities across major developed markets and industry-level data and to analyze them across industries.

Table 2 Environmental, social and corporate governance index statistics (last data available)

Country	Ner of firms	Environmental		Social		Corporate governance		Economic		Equal weighted	
		Mean	Median	Mean	Median	Mean	Median	Mean	Median	Mean	Median
Japan	430	0,639	0,828	0,504	0,533	0,157	0,116	0,391	0,333	0,417	0,419
Australia	295	0,359	**0,229**	0,353	**0,233**	0,580	0,604	0,411	0,329	0,393	0,286
Hong Kong	116	0,348	0,259	0,364	0,283	0,302	0,233	0,357	0,295	0,304	0,174
South Korea	63	0,572	0,664	0,505	0,461	0,160	0,101	0,485	0,517	0,435	0,482
Singapore	52	0,378	0,295	0,399	0,355	0,430	0,446	0,534	0,493	0,415	0,323
Taiwan	51	0,459	0,372	0,401	0,311	**0,141**	**0,047**	0,423	0,351	0,328	0,185
Malaysia	38	0,375	0,293	0,426	0,370	0,413	0,392	0,354	0,277	0,358	0,318
Total Asia/pac.	*1045*	*,492****	***0,436***	***,433*****	***0,331***	***,315*****	***0,226***	***,406*****	***0,343***	***,392*****	***0,312***
		(1, 2)		***(1, 4)***		***(1, 2)***		***(1,2,4)***		***(1,2,4)***	
China	60	**0,336**	0,240	**0,364**	0,289	0,226	0,180	**0,343**	**0,224**	**0,277**	**0,138**
Brazil	48	0,545	0,599	0,694	0,849	0,292	0,240	0,619	0,754	0,555	0,646
South Africa	46	0,645	0,742	0,776	0,848	0,598	0,618	0,689	0,762	0,729	0,817
India	44	0,547	0,538	0,628	0,623	0,310	0,276	0,605	0,663	0,524	0,508
Russian Fed	33	0,382	0,259	0,535	0,571	0,302	0,218	0,417	0,354	0,381	0,334
Total Brics	*231*	***,488*****	*0,484*	*,589****	*0,649*	*,341****	*0,279*	*,530****	*0,548*	*,487****	*0,5*
		(1, 2)		*(1,2,3)*		*(1, 2)*		*(1, 3)*		*(1, 3)*	
United Kingdom	376	0,601	0,700	0,656	0,730	**0,735**	**0,788**	0,567	0,621	0,684	0,783
France	98	**0,766**	**0,899**	**0,796**	0,894	0,639	0,692	**0,708**	0,761	**0,787**	**0,871**
Germany	84	0,699	0,873	0,720	0,828	0,369	0,354	0,640	0,682	0,645	0,730
Switzerland	67	0,580	0,655	0,566	0,596	0,526	0,597	0,624	0,756	0,592	0,740
Italy	54	0,502	0,445	0,667	0,781	0,487	0,451	0,598	0,619	0,572	0,582
Spain	54	0,702	0,874	0,795	**0,939**	0,552	0,627	0,658	0,825	0,728	0,858
Sweden	54	0,657	0,838	0,668	0,742	0,663	0,709	0,617	0,747	0,702	0,795
Netherlands	41	0,595	0,720	0,744	0,812	0,662	0,701	0,705	**0,786**	0,760	0,852
Total Europe	*828*	***,632*****	***0,766***	***,687*****	***0,782***	*,633****	*0,700*	***,614*****	***0,693***	***,685*****	***0,808***
		(2,3,4)		***(2,3,4)***		*(2,3,4)*		***(2,3,4)***		***(2,3,4)***	
United States	1127	0,409	0,294	0,443	0,393	0,724	0,763	0,520	0,541	0,526	0,492
Canada	291	0,363	0,238	0,367	0,300	0,730	0,773	0,438	0,377	0,459	0,409
Total N. America	*1418*	*,399****	*0,278*	*,427****	*0,370*	***,726*****	***0,766***	*,503****	*0,512*	*,512****	*0,471*
		(1,3,4)		*(1, 4)*		***(1,3,4)***		*(1, 3)*		*(1, 3)*	

*** significant at p-0.001 level
Numbers in parentheses indicate the group numbers from which this group was significantly different at p-0.001 level according to the Scheffe's pairwise comparison procedure. 1 = Europe, 2 = North America, 3 = Asia/Pacific, 4 = Brics

Table 3 Environmental, social and corporate governance index statistics (last data available)

Industries	Environmental			Social		Corporate governance		Economic		Equal weighted	
	N	Mean	Med.	Mean	Med.	Mean	Med.	Mean	Med.	Mean	Med.
Containers/Packaging	22	0,747	0,819	0,588	0,586	**0,738**	**0,840**	0,639	0,627	0,723	0,749
Utilities—Multiline	29	0,647	0,743	0,620	0,683	0,686	0,796	0,637	0,646	0,684	**0,851**
Paper/Forest products	31	0,758	0,828	0,641	0,702	0,694	0,795	0,540	0,574	0,694	0,668
Semiconductors/Semiconductor equipment	71	0,610	0,712	0,528	0,510	0,596	0,778	0,549	0,527	0,587	0,672
Commercial services/Supplies	108	0,495	0,486	0,539	0,565	0,691	0,761	0,567	0,592	0,591	0,617
Energy related equipment/Services	82	0,382	0,284	0,474	0,434	0,697	0,756	0,539	0,552	0,522	0,461
Aerospace/Defense	37	0,682	0,806	0,661	0,742	0,716	0,754	0,706	0,768	**0,738**	0,838
Healthcare equipment/Supplies	58	0,430	0,309	0,473	0,477	0,590	0,731	0,575	0,644	0,533	0,572
Homebuilding/Construction supplies	47	0,699	0,869	0,585	0,716	0,608	0,722	0,470	0,454	0,615	0,726
Communications equipment	30	0,699	0,798	0,568	0,632	0,653	0,715	0,586	0,687	0,653	0,694
Investment trusts	7	**0,221**	**0,113**	0,257	**0,168**	0,526	0,707	**0,055**	**0,046**	0,234	**0,157**
Biotechnology/Medical research	32	0,342	0,131	0,448	0,366	0,662	0,702	0,400	0,380	0,438	0,295
Oil/Gas	200	0,445	0,320	0,456	0,377	0,612	0,701	0,444	0,420	0,480	0,417
Air freight/Courier services	10	0,519	0,571	0,616	0,637	0,593	0,691	**0,736**	**0,831**	0,655	0,712
Hotels/Entertainment services	89	0,411	0,347	0,475	0,458	0,589	0,691	0,496	0,512	0,495	0,510
Software/IT services	115	0,394	0,268	0,443	0,387	0,590	0,685	0,549	0,555	0,483	0,427
Personal/Household products/Services	53	0,488	0,424	0,558	0,550	0,572	0,669	0,570	0,632	0,558	0,584
Utilities—water/Others	10	0,591	0,708	0,640	**0,904**	0,615	0,666	0,683	0,790	0,651	0,843
Media/Publishing	105	0,370	0,307	0,424	0,381	0,552	0,661	0,465	0,418	0,439	0,371
Healthcare providers/Services	41	0,232	0,119	0,351	0,307	0,617	0,658	0,441	0,406	0,375	0,303
Retailers—specialty	97	0,335	0,175	0,388	0,308	0,548	0,657	0,469	0,461	0,420	0,314
Insurance	151	0,411	0,290	0,450	0,414	0,583	0,650	0,585	0,610	0,500	0,452
Investment services	116	0,339	0,168	0,402	0,352	0,577	0,645	0,492	0,489	0,438	0,399

(continued)

Table 3 (continued).

Industries	Environmental			Social		Corporate governance		Economic		Equal weighted	
	N	Mean	Med.	Mean	Med.	Mean	Med.	Mean	Med.	Mean	Med.
Rails/Roads transportation	61	0,499	0,497	0,472	0,426	0,507	0,643	0,432	0,368	0,478	0,475
REIT—residential/Commercial	87	0,397	0,303	0,269	0,170	0,559	0,638	0,394	0,358	0,369	0,262
Beverages	44	0,618	0,734	0,596	0,760	0,514	0,626	0,553	0,634	0,591	0,783
Metal/Mining	248	0,449	0,387	0,453	0,391	0,560	0,619	0,387	0,312	0,447	0,410
Food/Tobacco	110	0,548	0,636	0,564	0,600	0,530	0,596	0,541	0,604	0,560	0,684
Biotechnology/Pharmaceuticals	82	0,498	0,504	0,543	0,575	0,518	0,585	0,489	0,500	0,515	0,494
Construction materials	33	0,623	0,770	0,569	0,587	0,530	0,582	0,459	0,472	0,556	0,581
Food/Drug retailing	48	0,537	0,657	0,545	0,592	0,520	0,562	0,574	0,660	0,571	0,709
Construction/Engineering	86	0,643	0,751	0,601	0,642	0,511	0,559	0,509	0,522	0,590	0,655
Machinery/Equipment/Components	177	0,675	0,818	0,581	0,642	0,491	0,549	0,576	0,653	0,605	0,681
Electric utilities	116	0,625	0,716	0,613	0,667	0,530	0,545	0,558	0,554	0,601	0,672
Household goods	36	0,708	0,892	0,623	0,761	0,534	0,536	0,535	0,562	0,634	0,758
Telecommunications services	119	0,484	0,423	0,564	0,620	0,483	0,529	0,592	0,604	0,533	0,525
Airline services	38	0,542	0,614	0,575	0,646	0,528	0,528	0,494	0,480	0,559	0,620
Chemicals	111	0,726	0,864	0,681	0,795	0,496	0,517	0,606	0,669	0,664	0,738
Financial services—diversified	7	0,439	0,390	0,449	0,589	0,437	0,508	0,462	0,474	0,455	0,578
Renewable energy	16	0,609	0,684	0,517	0,473	0,464	0,492	0,465	0,510	0,523	0,552
Coal	32	0,352	0,248	0,404	0,358	0,509	0,487	0,405	0,391	0,399	0,411
Banking services	291	0,435	0,289	0,507	0,461	0,450	0,464	0,516	0,508	0,464	0,393
Computers/Office equipment	39	0,757	0,901	0,680	**0,837**	0,489	0,459	0,606	0,640	0,676	0,790
Gas utilities	22	0,450	0,393	0,479	0,384	0,467	0,448	0,522	0,476	0,466	0,342
Diversified trading/Distributing	10	**0,796**	0,886	**0,707**	0,766	0,418	0,446	0,483	0,450	0,643	0,631
Retailers—diversified	48	0,445	0,347	0,451	0,354	0,465	0,432	0,499	0,490	0,455	0,390
Industrial conglomerates	50	0,565	0,618	0,527	0,503	0,441	0,432	0,466	0,412	0,497	0,441
Textiles/Apparel	34	0,550	0,617	0,602	0,693	0,401	0,403	0,559	0,628	0,543	0,618
Real estate operations	106	0,426	0,339	0,322	0,215	0,398	0,329	0,331	0,239	0,335	0,217
Leisure products	21	0,474	0,327	0,481	0,398	0,391	0,269	0,448	0,419	0,437	0,307
Marine services	27	0,525	0,585	0,513	0,479	**0,364**	0,243	0,501	0,464	0,466	0,455
Automobiles/Auto parts	79	0,751	**0,913**	0,611	0,757	0,381	**0,238**	0,560	0,587	0,602	0,685

The first snapshot shows that Asian/Pacific markets, in general, China, Hong Kong, and Taiwan in particular, are ESG laggards. On the other hand, European countries are way above the others in EGS scores, especially France, Spain, and the Nether-lands. In addition, countries in North America and in the BRICS group seem to have an intermediate position.

After the identification of the areas, a one-way ANOVA is conducted to assess the differences across the groups. Scheffe's pairwise comparison procedure is used to test for differences between individual pairs of groups if one-way ANOVA results are statistically significant.

Table 2 presents the summarized results for the four areas described earlier. One-way ANOVA results for each area were statistically significant at $p = 0.001$. Additionally, the follow up Scheffe's pairwise comparison results demonstrated that each area results were different from, at least, two other identified areas.

At an industry level, Utilities, aerospace, computers, and beverages are among the best-in-class in the overall rating, while Investment Trusts and Real State are clearly the worst-in-class industries. We should also highlight the difference between manufacturing and service industries. Manufacturing industries are more likely to have higher scores than service industries.

3 Environmental Scores

Following the EGS levels, European countries show a level of corporate disclosure of environmental practices higher than other countries. Besides China, Australia, and Hong Kong, countries in North America have the lowest scores for environmental measures, which highlight a lower disclosure and/or lower adherence to environmental standards. Additionally, companies in the two largest economies show little interest in the environment. Such behavior is in line with that seen at a political level, where countries systematically rejected Kyoto protocols (Altomonte 2008; Fairchild 2008). In this regard, Canada which also first signed up to Kyoto, recently stepped aside. This we think is a significant issue, especially when considering such a big number of large firms that should be an example to smaller firms in the industry (Gadenne et al. 2009; Ferrari et al. 2010). Nevertheless, over and above this static over-view, a dynamic analysis is needed.

We analyzed the environmental ratings for each year to see the evolution (Fig. 1).

Figure 1 provides environmental scores per area, which shows the tendency of these areas in environmental related issues. As in the static analysis, European companies are the leaders in environmental disclosure and/or activities. Every single year, these average scores of European companies have been higher than the aver-age scores of companies in the other areas. BRICS score in 2006 is not very significant as only five companies were reported in the whole area.

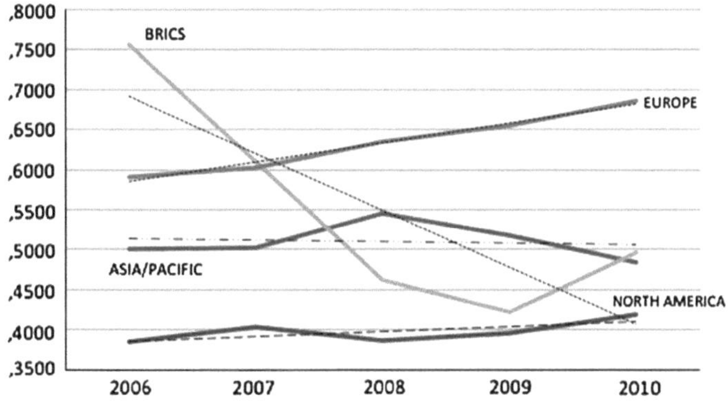

Fig. 1 Environmental scores evolution 2006–2010

We added regression lines to each of the set of points to show tendency (Martín Guzmán 1988). Tendency lines indicate a continuous improvement in European companies' environmental score. In terms of North American firms this improvement is held at a very slow rate. Asia/Pacific companies remained almost stable during the five-year period. BRICS should be explored from 2009 when a big number of companies had already been introduced in the analysis and there was a more stable sample to compare the evolution. Hence, BRICS scores seem to be growing, although we should wait a few more years in order to have more data to support this statement.

Sulaiman et al. (2002) looked for the interrelations between environmental disclosure, environmental performance, and economic performance. They found that "good" environmental performance is significantly associated with more extensive quantifiable environmental disclosures of specific pollution measures and occurrences. According to this statement and what several authors have found in support (Manescu 2010; Humphrey et al. 2012), we could expect that the higher the environmental scores (environmental disclosures), the higher the environmental performance, which states our hypothesis in this work.

As environmental scores show environmental practices relative to energy and water consumption (Mondéjar-Jiménez et al. 2011), we can expect companies with lower scores to consume more energy relative to their sales generation compared with other companies in the same sector with higher scores. Companies with high scores are more energy efficient or they use a higher amount of renewable energy sources. We can also expect a more efficient water consumption pattern in companies with higher scores. Therefore, for every dollar of revenue generated by environmentally oriented companies, their water and energy consumption is expected to be lower, revealing a more efficient usage of the resources. Also, waste recycling raises the degree to which a nation's corporations are environment friendly.

Greenhouse gas emissions are also included as a KPI in the Environmental score. We can expect Asia/Pacific, Chinese, and North American companies to be leading the production of greenhouse gas emissions. The higher emissions in Asia may be due to the high consumption pattern of their industries in an economy that is predominantly trade-dependent. The high emissions of North American companies can be due to the petroleum-based fuels dependency, electricity/power generation, and transportation needs. In comparison, European and Japanese companies excel in controlling green-house gas emissions.

If we look back, we can see that European governments have been promoting for a long time regulations in order to address these issues. Water consumption has been a challenge in Southern European Countries for decades, and fuel prices in Europe and Renewable energy promotion have made it more attractive for companies to seek energy efficiency and to invest in alternative energy sources. Finally, strong regulation and commitment has forced companies to reduce greenhouse emissions, waste and recycle practices and to promote eco-innovations (Segarra Oña et al. 2011; Segarra-Oña et al. 2012).

Clarkson et al. (2011) showed that positive (negative) changes in firms' financial resources in previous periods are followed by significant improvements (declines) in firms' relative environmental performance in the subsequent periods. Those findings mean that the financial performance of companies in growing markets like China or Russian Fed, or in well known profitable markets like North America are showing no or little evolution compared with European companies. This might lead us to think that higher forces are driving the environmental performance of the firms rather than a good financial performance (Porter and van der Linde 1995; King and Lenox 2001).

In addition, Clarkson et al. (2011) also found that significant improvements (declines) in environmental performance in the prior periods can lead to improvements (declines) in financial performance in the subsequent periods. However, Balabanis et al. (1998) found that firms' involvement in environmental protection activities was negatively correlated to subsequent financial performance. These differences may have been affected by the period when the studies where done. Nowadays, environmental issues are heavily present in the strategies of large corporations, which realize that it is affecting and/or will eventually affect their competitiveness.

Environmental scores by industry shows that, although manufacturing industries have significantly higher values ($p < 0.001$) than service industries, some services are taking the lead in environmental actions and disclosure. Industries with higher scores are also the industries with higher increase in environmental ratings over the pasts five years. Figure 2 reports the environmental scores evolution for the industries with the higher environmental disclosure and activities.

4 Social Scores

Social KPIs value the net employment creation by the companies, which we expect to be higher in those economies with rich resources that should be benefiting from emerging markets' demand and those economies with a higher growth in the GPD.

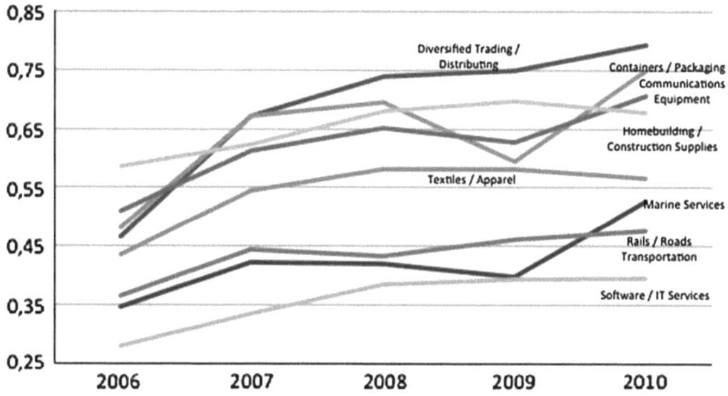

Fig. 2 Top environmental scores industries evolution 2006–2010

On the other hand, recession in the US and Europe will have increased the unemployment rate dramatically in Europe. Additionally, this situation may be affecting personnel turnover, which is a measure of better working conditions and job satisfaction, but can also reflect working culture. Employment security, new labor practices, flexible working hours, and labor unions are only a few elements that can affect this parameter. We expect, up to a certain point, that the lower the regulation, the more companies' working conditions and talent retention policy will affect personnel satisfaction and turnover. As a result, governments are playing an important role in setting the labor framework that allows employees to have enough freedom to look for a company that fits with their needs.

Top multinationals such as Unilever and Dupont consider health and safety as a very important performance indicator to benchmark their factories. It shows how employee-friendly and responsible companies are, and increases productivity and employee commitment to the company. Similarly, employee training and development affect companies' competitiveness and therefore countries' competitiveness. Nevertheless, this training should be focused on developing innovation skills that will increase the efficiency of the processes.

Furthermore, we can expect higher scores in those companies that take care of their employees; they care about health and safety and provide training, making them committed to the company and less likely to change to another firm.

Table 2 shows Spain, France, and the Netherlands as having more socially responsible companies. In terms of environmental aspects, European companies are leading the scores. We think this is because of a working culture focused on social benefits, on the one hand, and more mature organizational culture, on the other hand. This culture considers employees as the most valuable asset in the company and, therefore, searching to take care of this asset is the best possible objective. Only BRICS, regardless of China, have reached the same level as Europeans. Nevertheless, this is not being reflected in countries' social performance. Asia/Pacific and North American countries have a very small sense of social responsibility.

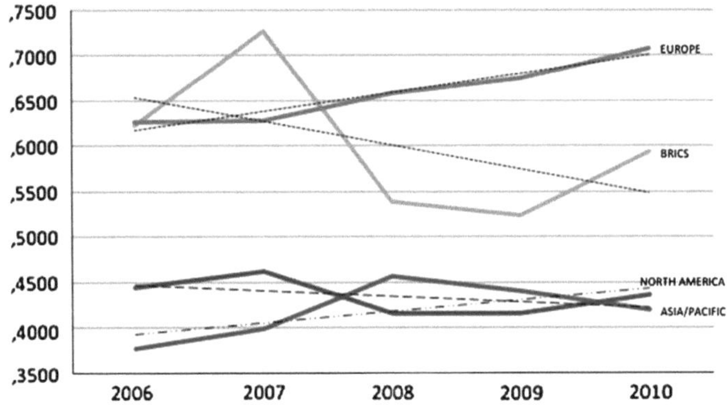

Fig. 3 Social score evolution by area

As regards evolution in the past years, we perceive similar tendencies as in the other two indicators. Except for China, and considering only the last two years, BRICS and European companies have clearly and constantly increased social responsibility. The United States seems to be in the same static situation, which is very disappointing as growing economies look to it as an example of prosperity. Thus, we can see that the regulation efforts of governments and company culture are driving again the evolution of the social scores (Fig. 3).

Social scores by industry reveal that industries closer to consumers and with higher social concern, automobile and chemical industries, have higher social scores. Other industries and services that are industrial suppliers and, therefore, far from the final consumer (mining or semiconductors) have less concern about social issues. This might be explained because of the impact of consumer behavior and its implications in the performance of the companies, which are felt differently in each industry. The evolution of social scores shows the industries with the higher increase in the last years. This figure indicates that some industries are feeling higher pressure to increase social concern and disclosure. Some industries might be forced to increase social ratings by consumers and others by their industrial clients, which might force the actors in the product or service value process to be more social (Fig. 4).

5 Corporate Governance Scores

Corporate Governance scores reflect the experience and the independence of the company's board of directors. The higher the score, the higher the experience and independence of the board member and the higher the participation of the shareholders. Governance score also accounts for board compensation and governance reporting.

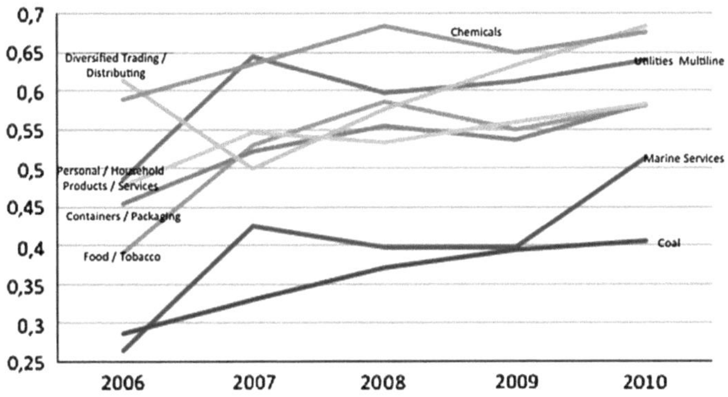

Fig. 4 Social score evolution by industry

We can expect higher board compensation and higher corporate governance disclosure in those firms with high scores.

When Europe is still dealing with the economic crisis, government has played a key role in raising ESG awareness. The European Commission and European countries are increasing efforts to strengthen transparency, ethics, and corporate governance through the promotion of new regulations.

Although the United Kingdom has the highest Corporate Governance score in Asset4 universe, US and Canadian companies are al-most at the same level.

The UK's independent regulator, the financial reporting council (FRC), has been promoting the Corporate Governance Code and high-quality corporate governance and reporting to foster in-vestments and to maintain the UK as an international financial center.

However, in this field, there is a huge gap between North American and European Companies and Asia/Pacific and BRICS companies. This aspect has not apparently affected the economic growth of companies based in these countries. We think it is because there is not a need for transparency when economic growth is important; governments will put aside this issue until other social is-sues in their countries are solved. Also, from the investors' point of view, high revenues are compensating the lack of information which, eventually, is just a measure of the risk.

We then look at the evolution over the past five years (see Fig. 3).

Again, the biggest effort to increase Corporate Governance standards has been made by European Companies. North American companies, as they are already in the top level, have maintained the level of Corporate Governance disclosure. BRICKS and Asia/Pacific companies are increasing their scores, but with such a low rate of in-crease, it will take a long time to reach the level of North American or European companies, unless local governments and market regulatory agents harden the regulatory framework and corporate governance codes for those firms (Figs. 5, 6).

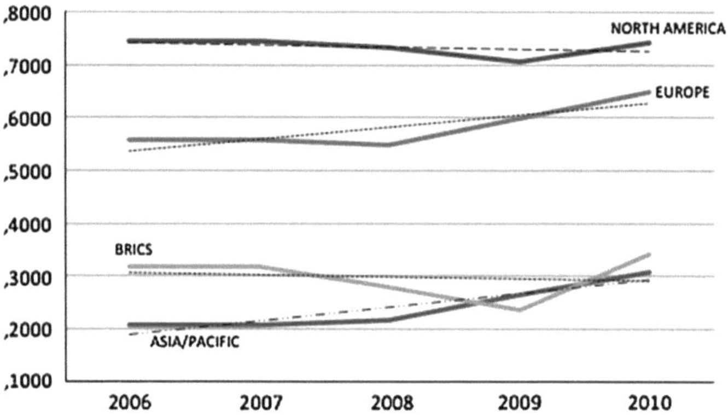

Fig. 5 Corporate governance score evolution by area

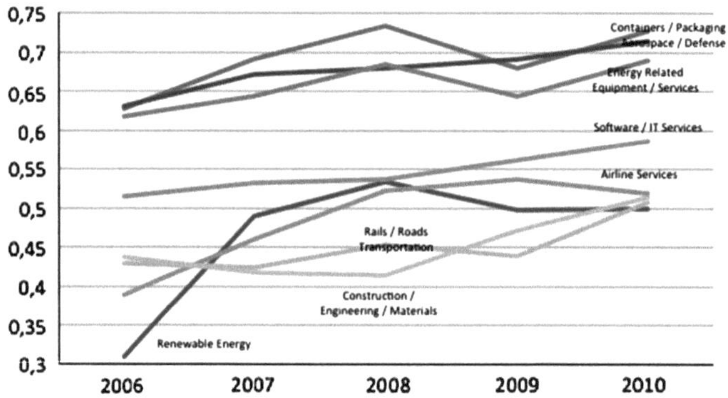

Fig. 6 Corporate governance score evolution by industry

Among industries packaging leads clearly the Corporate Governance indicators. We should underline the Automobile industry, which is at the bottom of the chart in this concept. No special characteristics of the industries lead the score in this concept. Looking at the evolution in the last few years, the industries that have been leading the growth in this indicator seem to be those with high growth rates like software and IT services, renewable energy or aerospace and defense. This evolution might be caused by the necessity to show more transparency to investors in order to improve financial position for the future.

6 Conclusions

Environmental, social, and corporate governance aspects are increasingly gaining attention from stakeholders, especially government, media, and investors. ESG exposure can be important at a company level, but also at a country level.

It is not surprising that Europe tops the list of largest ESG-conscious markets. European companies are a long way ahead of their Asian and American counterparts, heading ESG action. Europe, which has defined new stringent legislation, and ensured its enforcement among EU members, is addressing growing environmental challenges.

Asian markets, China in particular, have been laggards as evidenced by their low scores. The low rating, which normally implies lower disclosure, could be attributed to the lack of legal requirements to disclose ESG data. Although small companies tend to have lower scores than larger companies, aggregated data in each country certainly reflect the companies' tendency of the countries' firms, regardless of their size.

From another point of view, services industries are taking the lead over manufacturing industries in environmental aspects, while proximity to the consumer is leading social responsibility in industries and services. Corporate governance scores are growing and we expect them to grow in the future according to the financial needs of the industries.

Acknowledgments The authors would like to thank the Spanish Economy and Competitiveness Ministry for its support through the research project (EC02011-27369), and the Universitat Politècnica de València for its research funding to the project (PAID-06-2011-1879), "Innovative practices' on environmental firms "performance". Identifying moderating effects". Also the UPV Research Micro-cluster "Globalization, tourism and heritage: towards a sustainable development".

References

Altomonte S (2008) Climate change and architecture: mitigation and adaptation strategies for a sustainable development. J Sustain Dev 1(1):97–112

Balabanis G, Phillips HC, Lyall J (1998) Corporate social responsibility and economic performance in the top British companies: are they linked? Eur Bus Rev 98(1):25–44

Barba Sánchez V, Atienza Sahuquillo C (2008) La adopción de una estrategia ecológica en la empresa: aplicación de la teoría de recursos y capacidades, capacidades y dinámicas y responsabilidad social corporative (In Spanish). Revista de economía, sociedad, turismo y medio ambiente: RESTMA 7:31–53

Bassen A, Kovacs AM (2008) Environmental, social and governance key performance indicators from a capital market perspective. Zeitschrift für Wirtschafts- und Unternehmensethik 9(2):182–192

Besley T, Maitreesh G (2007) Retailing public goods: the economics of social responsibility. J Public Econ 91:1645–1663

Chatterji AK, Levine DI, Toffel MW (2009) How well do social ratings actually measure corporate social responsibility? J Econ Manage Strategy 18(1):125–169

Clarkson PM, Li Y, Richardson GD, Vasvari FP (2011) Does it really pay to be green? determinants and consequences of proactive environmental strategies. J Acc Public Policy 30:122–144

De-Miguel-Molina B, De-Miguel-Molina M, Albors-Garrigós J (2011) The innovative regional environment and the dynamics of its clusters. Eur Plann Stud 19(10):1713–1733

Delmas M, Vered DB (2010) Measuring corporate environmental performance: the trade-offs of sustainability ratings. Bus Strategy Environ 19:245–260

Duran D, Duran I, Diaconu A (2009) Issues on costs and investment for environmental protection. Environ Eng Manage J 8(4):705–708

Fairchild R (2008) The manufacturing sector's environmental motives: a game-theoretic analysis. J Bus Ethics 79(3):333–344

Ferrari G, Mondéjar-Jiménez J, Vargas-Vargas M (2010) Environmental sustainable management of small rural tourist enterprises. Int J Environ Res 4(3):407–414

Gadenne DL, Kennedy J, Mckeiver C (2009) An empirical study of environmental awareness and practices in SMEs. J Bus Ethics 84:45–63

Hong H, Kacperczyk M (2009) The price of sin: the effects of social norms on markets. J Financ Econ 93:15–36

Humphrey JE, Lee DD, Shen Y (2012) Does it cost to be sustainable? J Corp Finan 18:626–639

Kemp R, Parto S, Gibson R (2005) Governance for sustainable development: moving from theory to practice. Int J Sustain Dev 8(1):12–30

King A, Lenox M (2001) Does it really pay to be green? an empirical study of firm environmental and financial performance. J Ind Ecol 5(1):105–116

Kranjac M, Henny C, Sikimic U (2012) Do Europeans funds generate countries' sustainable development? Actual Probl Econ 5(31):386–396

Manescu C (2010) Stock returns in relation to environmental, social, and governance performance: mispricing or compensation for risk?. School of Business, Economics and Law at University of Gothenburg, Sweden

Martín Guzmán MP (1988) Métodos estadísticos en el análisis regional. Revista de Estudios Regionales 22:149–170

Melnyk SA, Sroufe RP, Calantone R (2003) Assessing the impact of environmental management systems on corporate and environmental performance. J Oper Manage 21(3):329–351

Mondéjar-Jiménez JA, Cordente-Rodríguez M, Meseguer-Santamaría ML, Gázquez-Abad JC (2011) Environmental behavior and water saving in spanish housing. Int J Environ Res 5(1):1–10

Porter ME, van der Linde C (1995) Toward a new conception of the environment-competitiveness relationship. J Econ Perspect 9(4):97–118

Segarra Oña M, Peiró Signes A, Miret Pastor L, Albors Garrigós J (2011) Eco-innovación, una evolución de la innovación? Análisis empírico en la industria cerámica española (In Spanish). Boletín de la Sociedad Española de Cerámica y Vidrio 50(5):253–260

Segarra-Oña M, Merello-Giménez P, Segura-Maroto M, Maroto-Alvarez, Peiro-Signes A (2012) Proactividad medioambiental en la empresa: clasificación empírica y determinación de aspectos clave (in Spanish). Tec Empresarial, 6(1)

Statman M, Glushkov D (2009) Equity investments: research sources investment theory: efficient market theory portfolio management: equity strategies. Financ Anal J 65(4), Available at SSRN: http://ssrn.com/abstract=1444355

Sulaiman M, Ahmad NN, Alwi N (2002) Management accounting practices in Malaysia: a survey of the industrial and consumer products sectors. Unpublished research report International Islamic University, Kuala Lumpur, Malaysia

Vargas-Vargas M, Meseguer-Santamaría ML, Mondéjar-Jiménez J, Mondéjar-Jiménez JA (2010) Environmental protection expenditure for companies: a Spanish regional analysis. Int J Environ Res 4(3):373–378

The Role of Maintenance in Reducing the Negative Impact of a Business on the Environment

Małgorzata Jasiulewicz-Kaczmarek and Przemysław Drożyner

Abstract New concepts of business management (Lean Manufacturing, Green Manufacturing or Sustainable Manufacturing) resulted in modified perception of maintenance. It is no longer a cost centre but a strategic business partner that plays a vital role that helps the organisation to achieve its eco-efficiency goals. Maintenance services have no direct impact on power consumption and other utilities or the amount of generated waste resulting from the applied manufacturing technologies. However, they may actively contribute to the reduction of environmental aspects identified in the organisation and to the improvement of its eco-efficiency. The two areas, maintenance and environment, are inter-dependable, both in terms of results of actions and the effects. What factors are in the responsibility of the maintenance? On what production system components can the activities of maintenance service affect positively?

Keywords Maintenance activities · Maintenance performance · Environmental maintenance BSC

1 Introduction

Industrial activity has been considered one of the major sources of environmental pollution, natural resources depletion, and natural environment degradation. Environmental impacts can be classified in many ways. With reference to the

M. Jasiulewicz-Kaczmarek (✉)
Poznan University of Technology, Poznań, Poland
e-mail: malgorzata.jasiulewicz-kaczmarek@put.poznan.pl

P. Drożyner
The University of Warmia and Mazury, Olsztyn, Poland
e-mail: przemyslaw.drozyner@uwm.edu.pl

M. G. Erechtchoukova et al. (eds.), *Sustainability Appraisal: Quantitative Methods and Mathematical Techniques for Environmental Performance Evaluation*, EcoProduction, DOI: 10.1007/978-3-642-32081-1_7,

range of impact it may be classified as local, regional or global; while as regards time—past, present, and future. One may also mention environmental impact referring to the range and time of activity, the phase of production, and the target of the impact. Environmental impact can affect water, air or soil. Organisations directly or indirectly extract fossil fuels, mineral resources, organic resources, water resources, etc., from the environment and input them into business processes as energy, raw materials, parts, products or water. Environmental burden such as the consumption of natural resources and modification of the state of the land is the result of these business activities.

The knowledge of the organisation's environmental impact is an essential element of a systematic and effective improvement of achievements in this respect (Hadas et al. 2011). Environmental policy, environmental objectives, and the relevant programmes are based on information referring to environmental aspects and impact of the organisation onto the environment. This gives the organisation an opportunity to prevent and limit its negative impact and to promote pro-environmental behaviour. Consequently, the challenge they face is to prepare a comprehensive quantity and quality analysis of raw materials and energy consumption as well as waste water, waste and contaminant emission to the atmosphere.

Assuming that an organisation is a set of manufacturing factors organised and coordinated in order to run a business connected with the production of goods and rendering of services, the following question occurs: in what way do individual manufacturing factors (an organisation's units or implemented processes) contribute to pursuing of the assumed strategy? What is their influence on achieving the assumed goals?

With no doubt, one of the major elements of an organisation's structure is maintenance. As irrespective of the area of business and supplied product, each organisation has technical resources (machines, equipment) that require maintaining in operation. By maintaining we wish them to performed tasks ordered by the user efficiently, i.e. with the optimum use of resources (materials, energy, etc.). In manufacturing context, maintenance management is the process of directing maintenance organisation effectively by utilizing administrative, human, financial, and material resources in an efficient and effective way through planning, scheduling, executing and monitoring their own progress for continuous improvement. Maintenance management's role is to provide support to production, and by providing reliable equipment and processes it helps organisation to be competitive and contribute to sustainable profitability; socially, economically and environmentally (Baluch et al. 2010).

New concepts of business management, such as Lean Manufacturing, new challenges relating to economical management of natural resources (Green Manufacturing) or Sustainable Manufacturing thinking resulted in modified perception of maintenance (Jasiulewicz-Kaczmarek 2013). It is no longer a cost centre but a strategic business partner that plays a vital role that helps the organisation to achieve its goals. Well adopted maintenance policy and availability of resources (human, material, technical, information-related) support organisations in achievement of their eco-efficiency goals in every phase of their equipment's life cycle.

Regular professional maintenance ensures the most eco-efficient use of equipment and the longest, cleanest life cycle, with the smallest environmental impact. It helps customers determine when and how to modernise equipment with thorough inspections of the safety, accessibility, reliability and energy efficiency.

Thus, inclusion of the maintenance function in the pro-environmental strategy is equally indispensible at the stage of investing in new machines and equipment and later during operation and abandonment phases (Napiórkowski and Szczyglak 2011; Napiórkowski et al. 2011).

A growing number of organisations uses different forms of defining its environmental impact, plans and takes pro-active steps to limit such an impact. Planning must be connected with an effective system of achievement assessment because evaluation is based on the adage "what gets measured, gets managed". For manufacturing processes, the availability of a set of indicators would allow comparing the environmental performance over time, highlighting optimisation potentials, deriving and pursuing environmental targets, identifying market chances, benchmarking against other companies or communicating results in environmental reports (Pawlewski and Borucki 2011).

Organisations use a variety of systems to evaluate achievements as well different measures and indicators (Greiner 2001; Feng and Joung 2009; Herva et al. 2011, ISO 14031 1999; ISO 14040 2006).

None of the above, however, refers comprehensively to maintenance as an important function performed in an organisation and its contribution to the implementation of its pro-active strategy.

The paper has been structured as follows. The first part identifies major causes of pro-active steps taken by organisations. Next, the maintenance system and its role in a product's life cycle are described. Attention is drawn to elements that are important in terms of environmental impact. The third part is an attempt to connect pro-environmental actions of maintenance with the objectives and strategy of the organisation. To do to so the Environmental Maintenance Balanced Scorecard (EMBSC) model is proposed. It allows examining of environmental issues with reference to maintenance from the perspective of strategy, setting out of goals as well as achievement assessment measures and indexes. Hence, it is possible to evaluate the quantitative contribution of maintenance to the implementation of the pro-active strategy of an organisation.

2 Company Motivators of Pro-environmental Activities

The evaluation of an organisation's environmental impact once again becomes a business priority. Clients, investors, legislators expect organisations to store data concerning their business' environmental impact and to reduce impacts from harmful factors. As a consequence, pro-environmental motivators of measures taken by organisations should be sought among market (clients, investors) and legal (legislators) requirements.

Market requirements formed by clients and investors often result from two factors. The first is reputation. They prefer to buy from and invest in organisations that care about ethical principles and meet all environment protection related requirements.

Cooperation with organisations that find it hard to meet legal environmental requirements or are locally perceived as "environmentally arduous or irresponsible" has a negative impact on image and may result in a drop of profit on product or service sales.

Another significant factor is the obligations assumed when adopting strategies aiming at sustainable growth. These obligations usually translate into requirements for all supply chain participants. Consequently, cooperation with large partners (e.g. automotive, power machines, furniture, food and other industries) is more and more dependent on the possibility of proving that the criteria of sustainable development have been met. Such criteria, industry-specific, are often set by organizations that dominate the market and groups thereof. Organizations are also more likely to undergo audits, independent verification and certification in individual areas of sustainable development (e.g., SA 8000, SMETA audit—Sedex Members Ethical Trade Audit, ISO 14001, GSCP—Environmental Reference Requirements 2010, World Business Council for Sustainable Development, Eco-efficiency indicators and reporting, etc.).

A factor that may also influence actions aimed at environmental protection are legal requirements.

According to the European Union requirements, legal acts pertaining to environmental protection impose onto organisations an obligation of preventing environmental hazards or limiting thereof to minimum. Reasonable use of natural environment by organisations is regulated by means of legal instruments such as ecological permits to use individual environmental elements and resources.

Based on the applicable acts and regulations and depending on the type and scale of their activities, organisations are obliged, among others, to hold permits regarding water-sewage management, protection of air against pollution, and noise protection.

As regards waste management, an organisation generating waste is obliged to keep a record of waste on specially prepared waste sheets. Identification and definition of the range of a manufacturing organisation's impact on natural environment elements is possible through permanent and comprehensive analysis of individual stages of the production process, technological operations and available infrastructure. This allows identifying the problems and needs of organisations with regard to environmental impact, indicating the possibility of elimination or reduction of the impact, and evaluating the effects of pro-environmental actions that have been taken.

As it results from a survey performed in 2010 by PBS DGA SA, the level of legal act binding is the first factor that differentiates the frequency of occurrence of a pro-environmental action. 88 % of organisations that indicated the level to be high or very high take pro-environmental actions. To compare, a corresponding percentage among organisations that indicated low level of binding with ecology-related legislation, or which are not bound by such acts, is 57 % (PBS DGA SA 2012).

Legislative regulations provide an impetus to change from non-sustainable to more environmentally friendly operations, but their influence should not be overestimated. The industry practice shows that environmentally friendly behaviour may be the answer to problems that companies face, namely: the increasing cost of energy, raw materials, and waste disposal (Golinska 2010).

Energy, raw materials and resources saving is the key determinant of technological progress and the aim of any entrepreneur wishing for profit. Nevertheless, such saving does not represent sustainable growth. Organisations must take different actions to combine, if possible, economic effectiveness and implementation of ecological responsibility rules. It is the best for them to seek such management methods that will form long-term basis for faster economic growth and promote eco-innovation and ecological safety. As a consequence, today's task for organisations is to create both economic and ecological values.

Implementation of the task results in seeking new, eco-effective technologies. A new challenge emerges for an organisation. It is changing the approach to technologies, processes, and products. This new approach forces new look onto the product through total environmental cost. Therefore, it refers to all participants of the product life cycle (manufacturer and client) and all life cycle stages, i.e. designing, manufacturing, using, maintaining, and abandoning/recycling.

3 Position of Maintenance in the Strategy of a Sustainable Development

A traditionally perceived scope of maintenance activities referred to manufacturing processes. Commonly accepted was the assumption that the main objective of maintenance function is to optimise availability of equipment at a minimum cost. However, shifting of the production paradigm towards a sustainable development resulted in maintenance paradigm change towards product life cycle management (Fig. 1), including maintenance in the chain of values of the

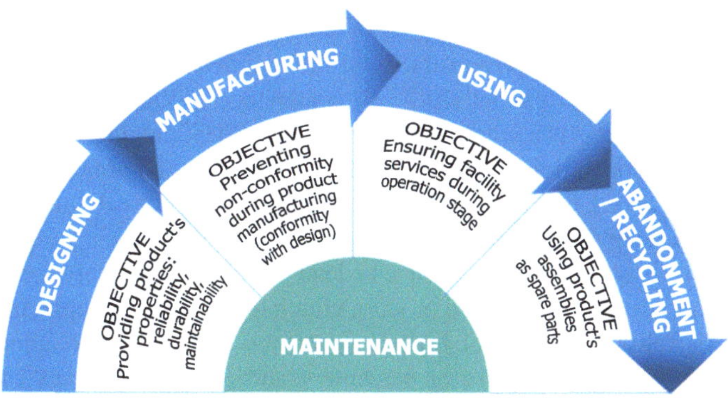

Fig. 1 Maintenance in product life cycle

entire organisation. To highlight and justify the new way of perceiving maintenance, Takata introduced the expression of 'maintenance values chain' (Takata et al. 2004).

Many decisions taken at the stage of designing, manufacturing and operation of a structure has a direct impact both on the effect and result in environmental and financial dimension and therefore must be adequately balanced (Johansson and Winroth 2010). For instance, a change of product structure or manufacturing method may cause a major reduction in energy consumption, amount of generated waste or environment pollution. The change itself may also significantly influence production economic results such as cost, productivity, quality and customer service. The essence is, therefore, to find a balance between ecological and economic benefits of the actions taken. This is a challenge to be faced by contemporary organisations both in terms of operational and strategic activities.

Currently, the tasks of technical services in most industrial organisations reach beyond the standard: framework of planning, completion and settlement of service and repair works. An important part of the attention of maintenance management is paid to rationalizing and optimising the decision making processes, both short- and long-term (Loska 2012), and in organisations applying the so-called good engineering practice, machinery maintenance is more than just a cost item to be avoided but first of all a proactive approach that may constitute an effective contribution to the organisation's growth and an integral part of green manufacturing.

Green manufacturing (GM) has been recognized worldwide as a key strategy for sustainable development and advanced model for manufacturing enterprises. The concept incorporates the principles of environmental protection and energy conservation into production and service activities to reduce industrial waste, save energy and scarce resource, and minimize pollutions to natural environment, while accomplishing production economy.

Maintenance services have no direct impact on power consumption and other utilities or the amount of generated waste resulting from the applied manufacturing technologies. However, they may actively contribute to the reduction of environmental aspects identified in the organisation and to the improvement of its eco-efficiency.

What factors are in the responsibility of the maintenance? On what production system components can the activities of maintenance service affect positively?

It turns out that this is a whole variety of possibilities, ranging from performing relatively simple operations and maintenance—repair and alignment or balancing, the use of advanced methods of technical diagnostics, correct lubrication, purchasing and strategy adopted to maintain the machines.

For example, shaft misalignment can lead to a 12 % increased energy consumption and incorrectly matched or worn clutch to a 4 % loss. Replacing the gear belt pulleys belts of traditional high performance enables a new generation of 2–4 % energy savings. The use of machinery and equipment drives energy-efficient bearings allows for 30 % reduction in friction and the machines can get 15 % higher speeds. The responsibility of maintenance services is also connected with choosing strategy of maintenance and correct strategy is directly linked to the mitigation of impacts on environment by ensuring the regular monitoring of performance of the machine by means of technical diagnostics tools, the elimination of major accidents and to prevent interruptions in the production cycle, focusing on the overall equipment effectiveness (OEE). In addition, maintenance is a process whose results can be seen in the measurable values and expenditure is relatively easy to manage. All this makes that to the businesses that apply the so-called 'good engineering practice' to maintain machinery is not only a cost to be avoided, but also an active action which could constitute an effective contribution to the development of the company.

The maintenance management has a variety of tools that enable participation of technical services of an organisation in all phases of a machine life cycle, and thus, participation in the performance of pro-environmental strategy of the organisation (Fig. 2).

The first stage of a product's life cycle is its design. Basic pro-environmental prerequisites of the designing phase of a technical object include implementation of the '3R' principle (reduce, reuse, recycle), and in particular: selection of structural materials with regard to environmental burden following their degradation, considering the possible reuse of the materials once the operation ends, ensuring high reliability of the machine during operation and possibly low power demand, ensuring the machine and equipment structures are durable, adjusting to repairs, easy diagnostics and service (Cempel et al. 2006).

Total Productive Maintenance (TPM) is a maintenance strategy developed to meet the new maintenance needs (Shetty et al. 2009; Ahuja and Khamba 2008). TPM is based on a "Zero-loss" concept with zero breakdown, accident and defects, to achieve high reliability, flexibility of equipment and reduce cost through minimizing wastage of manpower, raw material, energy, consumables, etc. One of the goals of TPM is to develop maintenance free equipment. One way to do this is to make improvements at the earliest possible stage, thus, at the stage

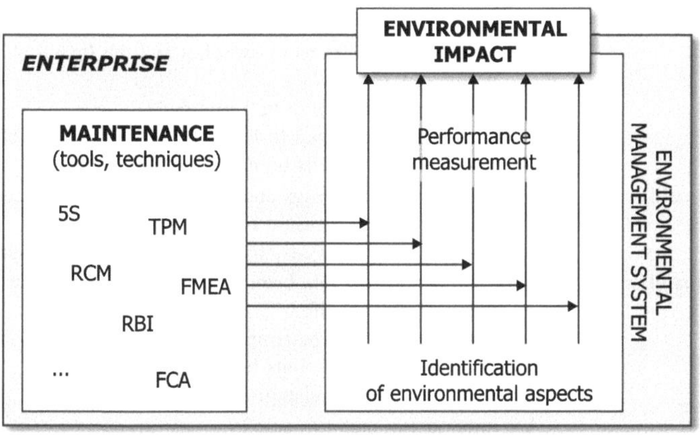

Fig. 2 Maintenance methods and techniques and their environmental impact

of designing. Achievement of the target is reached both by R&D units and engineering organisations. Engineering activity includes evaluation of the project with reference to the entire equipment life cycle cost, its weak points (Table 1). Therefore, benefits or losses may be defined that may result from defined operation and maintenance practice to ensure an ideal level of reliability, accessibility and ease of maintenance.

The next phase of a product's life cycle on which maintenance services have an impact and in which they have a proactive role is the machine's operation phase at the organisation. From the ecological point of view, maintenance of infrastructure in the phase of equipment operation is focused on ensuring systems, procedures and trainings that build the operational knowledge and skills as well as functional possibilities of the systems to prevent, manage and eliminate losses and environmental incidents.

A pro-environmental thinking should start at the operational level, i.e. from the machine and workplace perspective. The most important task for technical services and production personnel is to build a clean and well organised workplace. A solution that is most frequently used in organisations to build the 'cleanness culture' is the Japanese 5S practice. A 5S cornerstone is "the right thing in the right place at the right time"; anything else should be disposed of in a safe and environmentally correct manner. The 5S includes seiri (sort, organisation), seiton (set in order), seiso (shine, cleaning), seiketsu (standardize the cleaning), and shitsuke (sustain, discipline) and referred as the five keys to a total quality environment. The above are the key elements of the overall Management Operating System, including the elements that require managerial attention and whose supervision is only possible through sensor evaluation (e.g. sight, hear, smell).

Table 1 Examples of perspectives for equipment weak points searching

Facilitating autonomous maintenance	Can cleaning and inspection be easier?
	Can lubrication be centralized so that lubricant is supplied at just one or two inlets per equipment unit?
Increasing ease of operation	Can equipment be more resistant to operator errors, such as by changing the positions of switches and the layout of buttons on control panels?
	Can changeover procedure be simplified?
	Can standards be clarified to facilitate adjustments, or can measurement methods be made easier?
Improving quality	Have the precision settings and methods been determined (what to measure, how to measure it, limit values, etc.)?
	Is diagnostic equipment easy to set up? Does it have visual displays?
Improving maintainability	Have equipment life data been collected, and is work in progress to extend equipment life?
	Can parts replacement be simplified?
	Are self-diagnostic functions built into the equipment?
	Can oil supply and oil changing be simplified?
Safety	Are interlocking methods safe?
	Are there safety fences around hazardous equipment?

From the ecological perspective, 5S draws attention to all uncontrolled wastes and emissions that deviate from the standard, it fosters reasonable power consumption and gives a start to create and maintain standards for the working environment, introduces a change in culture. Table 2 presents examples of actions and potential benefits relating to 5S implementation.

In different organisations environment management and operational procedures are often independent of each other and are not available at workplaces. 5S practices assume a visualization of both the elaborated standards and maintenance performance. Visual management (laminated procedures at workplaces, performance boards, warning signs, etc.) may be used to improve designation of hazardous materials and waste and to improve the employee knowledge as regards the appropriate proceedings in terms of waste and emergency situations (Fig. 3). The combination of visual impulses, 5S practices, and operational procedures gives the employees a real chance of proceeding as per the applicable standards and improving of the environmental management.

In the phase of a technical structure operation in an organisation, the efficiency of maintenance actions depends on good cooperation between the machine and equipment operators and technical services staff. The cooperation is possible on the condition that both parties understand their own tasks as well as the task of the opposite party.

In real world the relation between the production and maintenance are regulated by the culture of mutual guilt. In case of failure it is the production that blames maintenance for indolence during the repair of damaged equipment, incorrect repair performance, lack of coordination of preventive actions with the production, etc. Whereas the maintenance blames the production for improper use of equipment, failure to report any damage symptoms leading to failing by the technical service to prevent such damages, etc. Such behaviour results in misunderstanding and conflicts whose tangible indicator is an increased failure frequency, low efficiency of use of the production equipment, low timeliness of repairs, large quantity of raw material waste (quality non-conformities of the products), energy loss, etc.

Breaking of the vicious circle (mutual blaming of the parties) is a prerequisite to carry out the strategy of maintenance, which on the one hand maximizes the availability and efficiency of equipment, controls the pace of equipment degradation

Table 2 Potential benefits of 5S

5S practice	Potential benefits
Sort	Improved use of space, reduced stock, reduced cost and more efficient work.
Set in order	Shortening of the task completion time, easier access, less errors, improved safety.
Shine	Clean workplaces and equipment, improved efficiency of machine and equipment operation, improved working environment and its surroundings, reduced amount of pollution and dirt.
Standardize	Improved working conditions and its quality, improved safety, reduced number of non-conformities (during task completion).
Sustain	Increased productivity and creativity, aiming at improvement and perfecting, adherence to agreements.

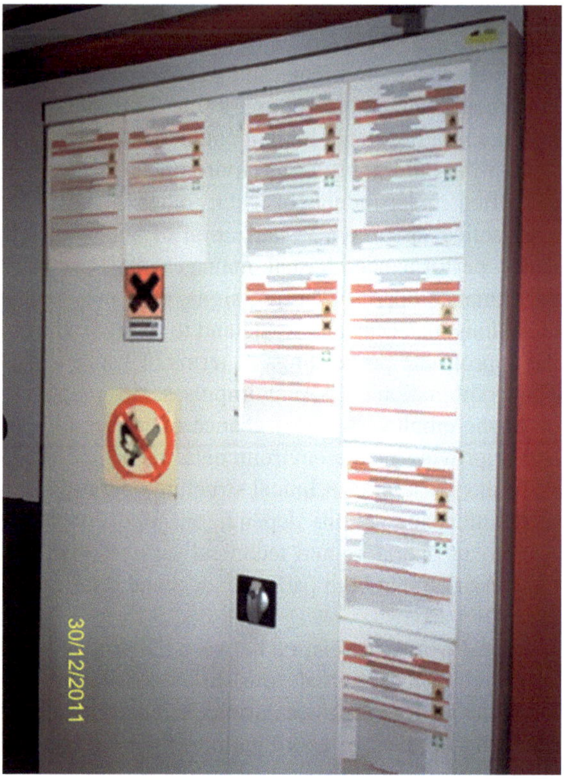

Fig. 3 Proper storing of dangerous chemicals

processes, and ensures environment safe and friendly actions, minimizing the total cost of operation on the other. Solutions may be found in teamwork and training whose objective is to provide knowledge and skills for operators and to include them in proactive preventive maintenance actions.

Preventive maintenance of machine and equipment is related to carrying out of a series of actions before a failure occurs. Its effectiveness depends not only on the quality of works performed by technical personnel but also on the speed and quality of information provided by the operators. Early identification of anomalies in machine operation is a necessary skill that should be built. The skill includes:

- first: ability to set the machine's parameters and conditions of work (to know how to tell normal conditions from abnormal),
- second: ability to maintain the machine's parameters and conditions of work (to know how to ensure satisfaction of normal conditions),
- third: ability to bring back the machine's parameters and conditions of work (to know when to react to abnormalities).

A program that builds such skills is the 'autonomous maintenance' carried out in organisations in relation with the implementation of the TPM. The program assumes providing operators with knowledge on the structure and principles of operation of the equipment they use and building basic skills for correct machine operation evaluation as well as performance of basic operation that is traditionally perceived as technical personnel task.

When introducing the concept of autonomous maintenance of machines and equipment, it is recommended to use the seven steps method (Table 3).

The aim of autonomous maintenance by the operators is:

First: stability of working conditions for the equipment and stopping the equipment degradation process, forming of the operator skills to perform daily routines (e.g. cleaning, lubricating) and small repairs,

Second: providing operators with knowledge on the equipment which they operate, possible problems that may occur and their reasons, as well as preventing such problems through early identification and elimination,

Third: preparing operators to proactive partnership and participation in programs (projects) of the equipment effectiveness and reliability improvement.

Including operators in the works pertaining to maintenance and handing over of responsibility and licenses to them enables better use of the knowledge of the equipment they hold, strengthens the feeling of self-esteem and enables aware participation in the organisation's target achieving.

Nevertheless, irrespective of how well the operators and technical personnel are prepared to observe current operation of the machine and identify deviations from standard conditions, equipment failures will happen and consequences thereof will pose a threat to maintenance and production quality, natural environment, human health and safety. Therefore, in the phase of operation it is necessary to have the knowledge of the machine functional failure's environmental impact and the selection of adequate operation actions and the condition monitoring system.

The methods of maintenance planning based on the identification and assessment of risk relating to equipment failure, enable planning of the production equipment maintenance in appropriate context (Narayan 2012).

Maintenance framework based on the risk assessment covers two main activities: risk assessment and technical service planning based the risk. The main objective of such activities is to reduce the overall risk that may lead to unpredictable failures. Priorities of operational actions (inspection, maintenance) are defined according to the quantitative analysis of a risk caused by failing sets of machines so that the total risk is minimized. The sets which as a result of the analysis have been classified as high risk are inspected and maintained more frequently and more accurately to ensure keeping of the allowed risk level. Organisations are more likely to promote maintenance in accordance with risk assessment. Such an approach enables defining of the appropriate proportions among individual maintenance policies, considering not only financial but also environmental and social issues (human safety).

The most frequently applied method in this area is the Reliability Centered Maintenance—RCM (Moubray 1995; Crocker and Kumar 2000; Mokashi et al. 2002; Niu et al. 2010). In the RCM analytical process all functions of any technical system, errors in performing these functions (damages) and all potential reasons

Table 3 Seven steps of "autonomous maintenance" introduction

Steps	Goals for equipment
Conduct initial cleaning	Eliminate environmental causes of deterioration, such as dust and dirt; prevent accelerated deterioration. Eliminate dust and dirt; improve quality of inspection and repairs and reduce time required. Discover and treat hidden defects.
Eliminate sources of contamination and inaccessible areas	Increase inherent reliability of equipment by preventing dust and other contaminants from adhering and accumulating. Enhance maintainability by improving cleaning and lubricating.
Develop cleaning and lubrication	Maintain basic equipment conditions (deterioration-preventing activities) cleaning, lubrication, and inspection.
Conduct general inspection skills training	Visually inspect major parts of the equipment; restore deterioration; enhance reliability. Facilitate inspection through innovative methods, such as serial number plates, colour instruction labels, thermo tape gauges and indicators, see through covers etc.
Conduct inspection autonomously	Maintain optimal equipment conditions once deterioration is restored through general inspection. Use innovative visual control systems to make cleaning lubrication/inspection more effective. Review equipment and human factors; clarify abnormal conditions. Implement improvement to make operation easier.
Organise and manage the workplace	Review and improve plant layout etc. Standardize control of work-in-process defective products, dies, jigs, tools measuring instruments, material handling equipment, aisles, etc. Implement visual control systems throughout the workplace.
Carry out ongoing autonomous maintenance and advanced improvement activities	Collect and analyze various types of data; improve equipment to increase reliability, maintainability and ease of operation. Pinpoint weaknesses in equipment based on analysis of data, implement improvement plans to lengthen equipment life span and inspection cycles.

for damage are being systematically identified, then direct effects of the above are identified, and finally, significance and consequences thereof (Fig. 4).

In the evaluation of consequences the RCM assigns any damage to one of the four categories that should be considered when elaborating the logical, decisive diagrams ('trees'): hidden failures that may lead to multiple failures and in extreme cases to

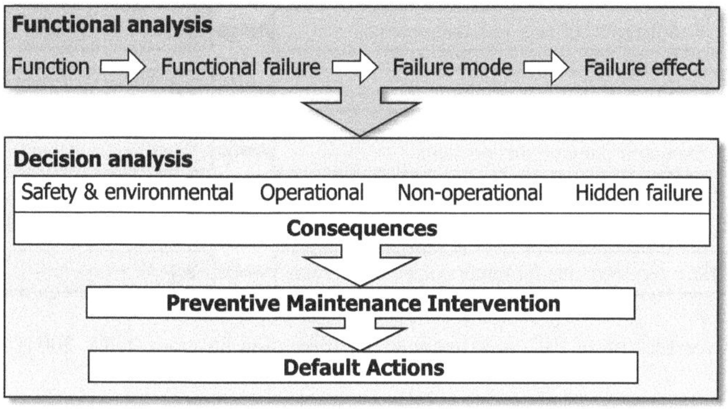

Fig. 4 The RCM process (Hipkin and Cock 2000)

catastrophic results, and open failures having an impact of human and/or environment safety, operational activity, and those which have no impact on operational activity (after a series of ecological catastrophes in the 1980s, drastic pro-ecological restrictions were introduced against business organisations in many countries; ecological consequences have gained appropriate importance in the RCM method). When all the information is provided, depending on the critical level of the failure, the most relevant maintenance policy is applied for the analyzed technical structure and specification of operational activities is created—whether preventive or not. These activities form a program of maintenance of the structure in a desired functionality adjusted to required operational parameters. Unlike other maintenance methods, the RCM allows for all options of activity: caused by diagnosed condition of the equipment, scheduled maintenance, scheduled component replacement, searching the hidden failures and single modifications (re-designing of components, change of operational procedures, additional operator training or other activities beyond the traditional scope of service works). One of the original conclusions arising from the analysis is also intentional allowing the failure to occur.

Risk based inspection (RBI) is a methodology which aims at establishing an inspection programme based on the aspects of probability and consequence of a failure. Inspection is carried out to reveal and confirm whether the process of degradation in a component is occurring. Inspection of the equipment will also give vital information on how the real process is developing compared to the expected scenario. This information can be used to define new measures to improve both the design of the equipment and the actions that are taken to preserve the risk level of the component. By conducting an RBI analysis the final results should answer what to inspect, when to inspect, where to inspect, how to inspect and what to report (Fig. 5).

The RCM process uses Failure Mode and Effect Analysis (FMEA). It is a deductive technique that consists in identification of failures at different levels of the machine structure complexity, their reasons and consequences to the entire system. The FMEA is focused on assessing the risks of Occupational health and

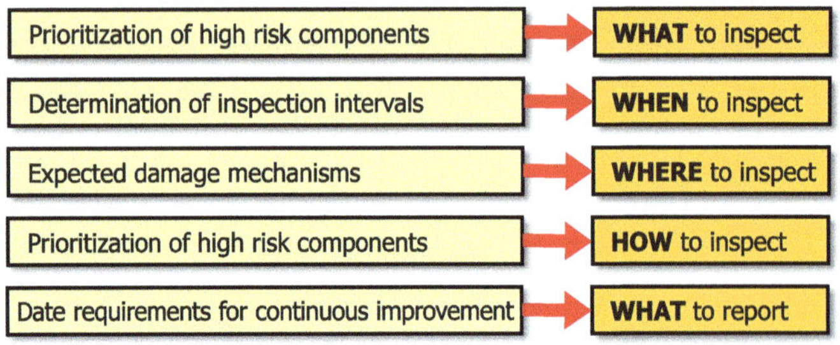

Fig. 5 Deliverables of an RBI assessment to the inspection program (DNV 2009)

safety (OHS), environment and quality management, which is based on three aspects, including "occurrence of failure" (indicating risk/probability that failure mode will occur as a result of a specific cause), "severity" (referring to an assessment of the seriousness of the effect of the potential failure mode in the process when it has occurred), and "detection" (referring to the probability that a potential failure will be detected).

FMEA analysis may be used both for the process of machine and equipment designing and for operation of machines and equipment and operation thereof, where based on historical data on the machine performance, analysis of the use environment, currently applicable legal requirements, it allows to identify possible non-conformities, their grounds and effects and to select appropriate preventing actions. In the context of environmental management the method will enable:

- systematic testing of internal environment protection and legal requirements,
- focusing on the most important activities to improve the condition of natural environment,
- delivering and influencing the most important environmental aspects,
- making aware environmental impact implementation easier.

Another tool employed to analyse reliability and safety of a system is fault-tree analysis (FTA). It provides an objective basis for analyzing system design, justifying system changes, performing trade-off studies, analyzing common failure modes, and demonstrating compliance with safety and environment requirements. It is different from a failure mode and effect analysis in that it is restricted to identifying system elements and events that lead to one particular undesired event.

Many reliability techniques are inductive and concerned primarily with ensuring that hardware accomplishes its intended functions. Fault-tree analysis is a detailed deductive analysis that usually requires considerable information about the system. It ensures that all critical aspects of a system are identified and controlled. This method represents graphically the Boolean logic associated with a particular system failure. Fault-tree analysis provides options for performing qualitative and quantitative reliability analysis. It helps the analyst understand system failures deductively and points out the aspects of a system that are important with respect to the failure of interest.

The universal objective of the maintenance process is to use the knowledge of abnormalities in machine operation and of incidents (both potential and occurring) in order to achieve optimum safety, from the point of view of people and environment, at the lowest possible cost. In industrial practice this transforms to optimisation of decision making processes concerning planning and completion of functional and servicing-and-repair works. Planning of technical service at a high level may be successful if it is based on reliable data from the operation level. This makes monitoring of the production equipment condition a key factor to support a balanced production.

A majority of machines may be monitored on continuous basis without stopping. Permanent monitoring and diagnostic systems ensure high stability and repetitiveness of the measuring process and allow uninterrupted analysis of measured values and detection of any excess of the preset limits. One of the important advantages of the continuous systems is the possibility of integration with control systems of the monitored machine, central control system, and visualization of the measured values in Supervisory Control and Data Acquisition (SCADA) systems or a superior, specialized diagnostic software. SCADA systems are currently standard elements of the machine monitoring systems. Their basic task is to acquire and visualize measuring data that describe current condition of the monitored machine and to generate warnings and alarms whenever machine working parameters exceed the preset levels. In case of failure, SCADA, as a superior system, offers a possibility of stopping the machine, its part or the entire technological line to minimize the failure spreading. Filing of measuring data—both during regular work and in alarm situations—is, however, from the maintenance perspective, only 'a raw material' that should be processed to get the valuable information. This processing is performed by specialized diagnostic software. An analysis of collected data concerning equipment operation parameters in combination with working conditions information (start-up, idle run, overload, etc.) requires extensive calculation expenditures such as frequency, modal analysis or modelling. With such a tool it is possible to relatively precisely define the condition of a given element or even define which failure of a given part may be expected or had already occurred. A well prepared monitoring system allows precise determination of the time a machine is withdrawn from operation, thus, eliminating raw material (non-conformant products), and energy losses and extending to the maximum the machines operation time, resulting in tangible financial benefits.

Another tool in the hands of technical personnel are diagnostic and forecasting test methods (Mikołajczak 2011) e.g. vibro-acoustics, vibration measuring, thermo-vision, oil analysis. Technical diagnostics, next to tribology, reliability, safety theorem and operation theorem, is one of the basic sciences on rational use of objects (Żółtowski 2008).

The objective of oil diagnostics is defining the usability of the used oil on the one hand, and checking whether is contains metallic particles that would prove wear of certain elements of the equipment, on the other (Szafrański 2011). From

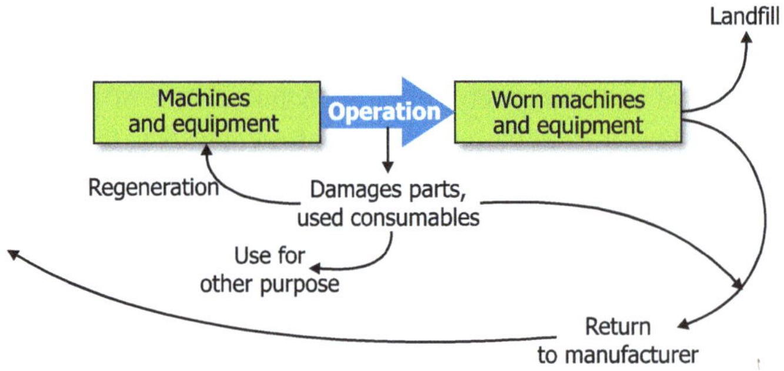

Fig. 6 'Abandonment' phase in a product's life cycle

the environmental perspective lubricants are non-renewable products. By maximizing their life span we reduce the organisation's negative environmental impact, minimize financial burden related to their disposal. Appropriately selected lubricant may have an impact on the increased level of equipment operation stability. Diagnostic tests of this area give only a small example of an organisation's possibilities to reduce environmental impact. Rational lubricating management should cover all activities, from oil selecting, through its storing, delivery to machine, its maintenance during operation, appropriate machine tooling (venting, drainage) adjusted to the conditions in the machine working environment, well designed oil analysis system, etc. As experiments show, improving actions in the area of lubricating management may lead to wear reduced by even 30 % (Jasiulewicz-Kaczmarek 2013).

Abandonment is the last phase of a machine's life cycle consisting in final withdrawal from operation. It is a condition when the machine reached its limit value of wear and further operation is impossible or uneconomical. The problem of managing of the worn parts and materials from the abandoned machine with the minimum burden to environment and certain economic result arises (Fig. 6).

Eco-design that has been developing for years is an approach where complex technical objects are designed so that recycling of materials, from multiple uses and reuse of elements (assemblies, parts) in several generations of machines for repair and modernisation is possible. Therefore, the object's abandonment phase requires taking steps resulting not only in transporting the machine to a landfill but also reuse of its elements. Each worn machine or equipment contains a series of valuable raw materials and subassemblies to be reused in repair and renovation of structurally similar technical objects. The task of maintenance personnel is to appropriately assess the fitness of assemblies and parts for other machines and equipment owned by the organisation and repairing or regenerating thereof for further use.

Despite their environmental arduousness, worn machines and equipment may constitute a source of valuable recyclables once reasonably recycled.

4 The Idea of Maintenance Performance Assessment from the Perspective of Environmental SD

Holistic approach to shaping maintenance system should include internal and external conditionings of a company, business strategy of the company and projection of how future changes in maintenance system will influence efficiency of company's functioning (Pawlowski et al. 2006).

All the aforesaid areas of activity of technical personnel must be covered by one coherent system of management where the goals, roles and tasks are defined and which is coherent with the strategy, needs and possibilities of an organisation.

The objectives of maintenance and the strategy of achievement thereof derive from the requirements of interested parties as well as goals and strategies of the organisation (Pinjala et al. 2006; Rosqvist et al. 2009), integrated at different levels of its organisational structure both top-down and bottom-up. The natural consequence is the need of designing and implementing a well-organised maintenance management system and a measuring system to assess its performance in terms of strategy. According to Alsyouf (2006) such a system should enable:

- evaluation of the maintenance function share in the organisation's business goals achieving;
- defining the strong and weak points of the implemented maintenance strategy;
- creation of stable base to develop a comprehensive strategy of maintenance improvement with the use of quantitative and qualitative data;
- re-analyzing and comparative analyzing of the implemented practices and maintenance performance with the best practices within and outside the same industry, and
- tracking the maintenance impact and indicating the relations between operational and financial activities (measures) in a holistic approach.

From a strategic point of view the process of maintenance management may be presented as a sequence of actions transforming the strategic goals into maintenance objectives and a strategy of achieving them, with a built-in system of performance metric that would enable assessment of efficiency and effectiveness thereof (Fig. 7).

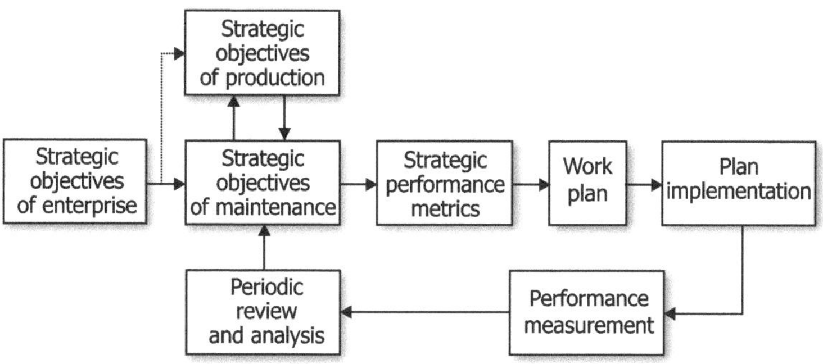

Fig. 7 Strategic maintenance-management process

A true challenge for maintenance management is to define the method of goals down-cascading, feedback up-aggregating, and integrating of activities performed by different internal organisation units of the organisation, so that the total effectiveness of maintenance and the desired business goals are achieved. This means that an organisation should define a hierarchy of goals where the maintenance objectives (e.g. 98 % availability of equipment) are means to achieve the overall organisation goals (e.g. production volume 5 tons/h).

Measuring of maintenance performance is an indispensable activity that may, for example, serve defining of the adjustment to new trends in equipment use and maintenance strategy, evaluation of employee safety and health caring, and meeting environmental challenges resulting from the organisation strategy.

Literature gives many examples of the concept of maintenance performance measuring. The Total Productive Maintenance concept (Nakajima 1988), launched in the 1980s, provided a quantitative metric called Overall Equipment Effectiveness (OEE) for measuring productivity of manufacturing equipment's (Jasiulewicz-Kaczmarek 2011). A hierarchical system of performance indicators of maintenance efficiency and the classification thereof referred to three main dimensions of maintenance performance (OEE, production costs and production quality) was proposed by Komonen (2002). Parida proposes a multi-criteria hierarchical framework for maintenance performance measurement (Parida and Chattopadhyay 2007) that consist of multi-criteria indicators for each level of management (i.e. strategic, tactical and operational). Al-Najjar (2007) proposes a model to describe and quantify the impact of maintenance on business's key competitive objectives related to production, quality and cost.

In 2007 the European standard for maintenance key performance indicators (EN: 15341 2007) was established to support the management in making the best use of the maintenance function in order to utilize all the technical assets in a more competitive way. The indicators can be used to measure the status, compare (internal and external benchmarks), diagnose (analysis of weaknesses and strengths), identify the objectives or goals, support to maintain and improve performance and continuously measure the evolution. The standard provides the management with a system of indicators to measure the maintenance performance considering the economical, technical and organisational aspects (Fig. 8). Maintenance Performance is the result of the utilization of resources in doing activities to provide actions to retain an item in, or restore it to, a state in which it can perform the required function. The Maintenance Performance is depending on influencing factors—external and internal, such as location, company culture, production process, strategies, policies and human competences—and is carried out implementing activities (example: corrective maintenance, preventive maintenance, improvements), using organisational methodologies (example: centralized, decentralized, outsourced, multiskilled, etc.), utilizing labour, information, materials, resources, tools and operating techniques.

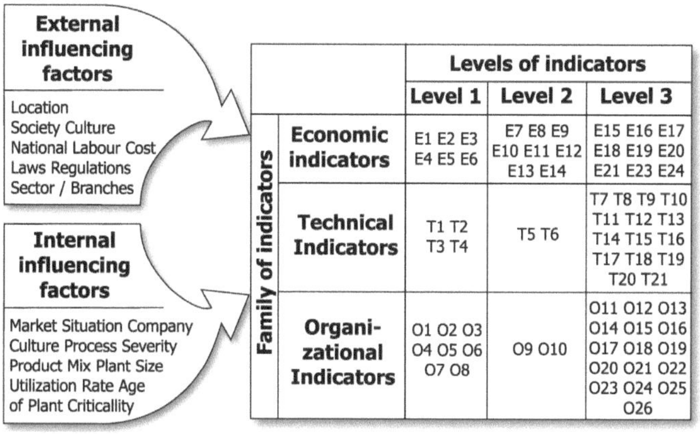

Fig. 8 Maintenance influencing factors and maintenance key performance indicators (EN: 15341 2007)

Exemplary "environmental" indicators of maintenance in this set are T12 and T13.

T12	$\frac{\text{Number of failures causing damages for environment}}{\text{Total number of failures}} \times 100$
T13	$\frac{\text{Number of failures causing potential damages for environment}}{\text{Total number of failures}} \times 100$

Many organisations find the set of indicators provided in the standard too complicated and incomprehensible. With reference to the environmental dimension of maintenance activities it is difficult to find a relation between the organisation goals and strategy and the maintenance indicators.

One of the recommended methods is the Balanced Scorecard (BSC), considered to be a balanced management system because it promotes equilibrium between short- and long-term objectives, between financial and non-financial measures, between indicators of tendency and occurrences, between internal and external perspectives of performance (Goncalves 2009). The balanced scorecard, developed by Kaplan and Norton (1992), is the most popular and balanced performance measurement framework, used by the most of industries all over the world (Quezad et al. 2009; Thakkar et al. 2007; Lawrie and Cobbold 2004; Akkermans and van Oorschot 2005). Unlike the traditional methods that focus on supervision, BSC is focused on the overall strategy and vision of the organisation and emphasizes achieving of target results.

Application of BSC gives an opportunity of better conformity between the goals worked out in the process of Environmental Management System and the organisation's strategy (Zingales et al. 2002). To obtain such a conformity the

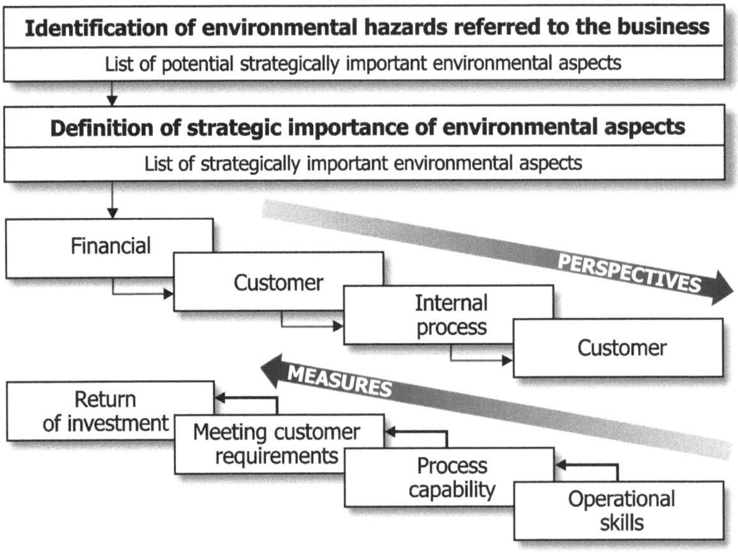

Fig. 9 The chain of cause and effect relationship from environmental BSC perspective

environmental aspects must be classified and integrated with the scorecard system depending on their strategic meaning, similarly to all potentially important strategic aspects (Fig. 9).

Environmental aspects can represent strategic core issues, for which lagging indicators have to be defined. These lagging indicators measure whether the strategic core requirements in the perspective have been achieved. Performance drivers as represented by leading indicators show *how* the results in each perspective, reflected by the lagging indicators, are to be achieved. Performance drivers are highly business specific but there are once again categories to support identification.

Figge et al. (2002) proposed to check systematically all pertinent environmental aspects by answering the following questions when going through the four conventional perspectives:

1. Does the environmental aspect represent a strategic core issue for the of our business unit (→ environmental lagging indicator)?
2. Does the environmental aspect contribute significantly to a strategic core issue and therefore represent a performance driver for the business strategy of our business unit (→ environmental leading indicator)?
3. What is the substantial contribution of the performance driver to the achievement of a strategic core issue?
4. Is the environmental aspect simply a hygienic factor, which necessarily has to be well managed but leads to no particular strategic or competitive advantage?

BSC is specific of organisations for which it was developed and allows building Key Performance Indicators to measure technical personnel management

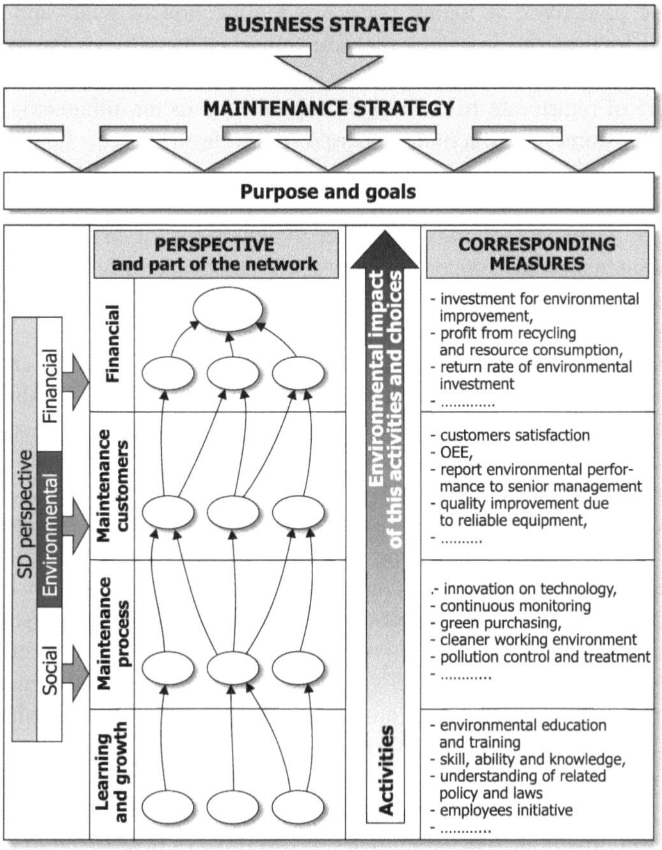

Fig. 10 Environmental Maintenance Balanced Scorecard (*EMBSC*)—a model concept

performance, referred to the organisation's strategic goals (Jasiulewicz-Kaczmarek 2012). The environmental dimension of maintenance may be described by each of the four Balanced Scorecard perspectives, for instance (Fig. 10):

- financial—meaning, for example, return of the capital invested in new diagnostic technologies,
- maintenance customers—meaning satisfaction of requirements and delivery of value mainly to internal clients (e.g. production),
- maintenance process—including but not limited to prevention of failures, identifying and implementing modifications to production equipment, management of materials, power and waste as eco-efficiently as possible,
- organisational (learning and growth)—meaning creating a new culture that values knowledge and skills, involvement in improving actions, which is reflected in maintenance, production, logistics, etc., employee daily choices and actions.

Strategic objectives of maintenance are a collection of goals and each new perspective focuses on processes that are important to achieve the goals determined in the previous perspective. In the financial perspective the stress is put on the measure of return rate from invested capital. The factor influencing the above measure is customer satisfaction expressed by the level thereof. Internal customers' satisfaction is related with cooperation conditions offered by technical personnel and the quality of services they render. All actions and decisions taken by maintenance management must conform to the needs of maintenance internal clients (maintenance is a supportive process). To react adequately to the needs and expectation of its customers, maintenance personnel must improve the parameters of service and method of delivery thereof.

The maintenance customers' perspective contains two measure groups: area activity results measure and measure of factors determining the value for customers. Service value deterring measures may be divided into service attributes (e.g. quality) and customer relations (e.g. time of service completion).

The internal processes perspective is a logical supplement of the customer perspective as regards maintenance activity assessment. In this perspective the key activities and processes influencing achievement of objectives set in previous perspectives are identified. Internal processes are assessed in terms of whether they meet customer expectations. After identifying the internal processes that are crucial from the customer perspective a metering system may be identified.

The last but not least of the perspectives (learning and growth) covers objectives and measures that determine area in which perfecting is expected in order to reach a major improvement in performance. The learning and growth targets form a basis supporting the completion of goals covered by the three remaining perspectives. When analyzing the learning and growth perspective grounds on which the long-term growth is to be based must be determined. Investment in employees, IT technologies and appropriate organisational procedures are required for a long-term and stable development.

The EMBSC model presented in the figure is an attempt of showing the importance and contribution of maintenance in the pro-environmental strategy of an organisation. Maintenance is able to create new productivity and help saving raw materials and energy, protect environment and increase profits in the industrial production. The value of such actions is countable and must be measured both in terms of management assuming responsibility for achieving the goals they set and identifying appropriate actions as well as directing the effort of employees to such achievement.

5 Conclusion

Maintenance, similarly to other functional areas of an organisation, is under constant pressure of cost cutting, achievement reporting, and mission supporting. From the point of view management these seem to be reasonable expectations because maintenance, as an auxiliary process, plays an important role in the organisation's operation. However, looking onto the activities of technical services from the perspective of a product life cycle one may notice new opportunities of strengthening the organisation's eco-effectiveness.

Practically, each phase of a product's life cycle requires maintenance. When designing a technical structure, one must allow for its total and comprehensive environmental impact. It applies not only to the use of such a structure/building but also all actions relating to maintenance thereof: lubricating, maintenance and repair. At this stage it is the knowledge acquired from organisations using the structures that matters. By monitoring the operation of equipment and performing technical servicing, maintenance service of such an organisation is a source of information on the potential to increase environmental effectiveness and often initiates changes to the building structure.

Another phase of a product life cycle that may be influenced and is actively participated by maintenance service is the stage of machine operation. From the ecological point of view, maintenance of infrastructure traffic in this phase is mainly focused onto two aspects. Firstly, by ensuring systems, procedures, and training that build the operators' skills and operational knowledge. Secondly, by ensuring material, technical, IT, and financial resources. These actions aim at prevention, management, and elimination of losses and environmental incidents when using technical buildings for manufacturing purposes.

Having such a potential of pro-environmental possibilities of maintenance, it would be highly irresponsible on the part of the management to ignore them in the process of defining an organisation's target and strategy. Therefore, the need to integrate the objectives and strategies of maintenance with the environmental objectives and strategy is obvious. How to do it?

The tool proposed herein is EMBSC. Why this tool? If we look at the scope of relation between maintenance and other functional areas of an organisation and effects desired by management we will see as follows. Eco-effectiveness is desired and expected. It may be achieved by meeting the requirements of our internal (e.g., production) and external stakeholders (although for external stakeholders maintenance actions are not visible, while the results and effects thereof are perceptible). If one wants to meet the requirements, maintenance processes must be adequately organised and then well managed. This needs people's knowledge and involvement. Thus, moving from top to bottom, we have logically passed the path of all BSD perspectives. The bottom to top order looks similarly, but here we measure the results/effects being the natural consequences of measures/decisions that were taken. One may, therefore, form cause and result relations between actions and their results and quantify the values thereof. BSC also gives an opportunity to quantify as quantity, quality, finance and non-finance, offering another important function—communication. Technical language is usually incomprehensible for economists, while the economists' language seems to be difficult to understand by engineers. This inconvenience and possible cause of conflict may be eliminated.

What else could be achieved by using the tool? For many years maintenance has been perceived as a 'cost'. This has been partially true, as it is an auxiliary process. Therefore, from an external client's perspective maintenance does not add value. However, from an internal client's perspective, today maintenance is a business partner. Through provision of a reliable machinery and reduction of operational risk, maintenance generates internal value that may be included in financial indexes. EMBSC enables to understand the meaning of maintenance inside an organisation and the contribution it makes to increase the ecological effectiveness. The most often, reports on achievements are submitted to authorities and external supervision units and local personnel are not appropriately informed.

References

Ahuja IPS, Khamba JS (2008) Total productive maintenance: literature review and directions. Int J Qual Reliab Manage 25(7):709–756

Akkermans HA, van Oorschot KE (2005) Relevance assumed: a case study of balanced scorecard development using system dynamics. J Oper Res Soc 56(8):931–941

Al-Najjar B (2007) The lack of maintenance and not maintenance which costs: a model to describe and quantify the impact of vibration-based maintenance on company's business. Int J Prod Econ 107(1):260–273

Alsyouf I (2006) Measuring maintenance performance using a balanced scorecard approach. J Qual Maintenance Eng 12(2):133–149

Baluch N, Abdullah CSB, Mohtar SB (2010) Maintenance management performance: an overview towards evaluating Malaysian palm oil mill. Asian J Technol Manage 3(1):1–4

Cempel CZ, Kasprzak J, Kłos Z (2006) Ekoinżynieria: ku holistycznemu projektowaniu i zarządzaniu środowiskiem, Ogólnopolska konferencja naukowa Zrównoważony rozwój w teorii ekonomii i praktyce, Wrocław

Crocker J, Kumar UD (2000) Age-related maintenance versus reliability centred maintenance: a case study on aero-engines. Reliab Eng Syst Saf 67:113–118

Det Norske Veritas (2009) Recommended practice DNV-RP-G 101: risk based inspection of offshore topside static mechanical equipment. Det Norske Veritas, Høvik

EN: 15341 (2007) Maintenance: maintenance key performance indicator

Feng SC, Joung CB (2009) An overview of a proposed measurement infrastructure for sustainable manufacturing. In: Proceedings of the 7th global conference on sustainable manufacturing. http://www.nist.gov/customcf/get_pdf.cfm?pub_id=904166. Accessed 29 June 2012

Figge F, Hahn T, Schaltegger S, Wagner M (2002) The sustainability balanced scorecard: linking sustainability management to business strategy. Bus Strategy Environ 11:269–284

Golinska P (2010) From Traditional Non-sustainable production to closed loop manufacturing: challenges for materials management based on PPC and EMIS integration. In: Teuteberg F, Marx Gomez C (eds) Corporate environmental management information systems: advancements and trends. Hershey, New York

Goncalves HS (2009) Proposal of a strategy model planning aligned to the balanced scorecard and the quality environments. TQM J 21(5):462–472

Greiner TJ (2001) Indicators of sustainable production: tracking progress. A case study on measuring eco-sustainability at Guilford of Maine, Inc 1. Greiner environmental, Lowell center for sustainable production. http://sustainableproduction.org/downloads/Guilford%20Case%20Study.pdf. Accessed 8 June 2012

GSCP: Environmental Reference Requirements (2010) http://www.oecd.org/site/agrfcn/48593472.pdf. Accessed 8 June 2012

Hadas L, Stachowiak A, Cyplik P (2011) Decision making model in integrated assessment of business-environment system: a case study. In: Golinska P et al (eds) Information technologies in environmental engineering. Springer, Berlin

Herva M, Franco A, Carrasco EF, Roca E (2011) Review of corporate environmental indicators. J Cleaner Prod 19:1687–1699

Hipkin IB, Cock CD (2000) TQM and BPR: lessons for maintenance management. Omega 28:277–292

ISO 14031 (1999) Environmental management: environmental performance evaluation: guidelines

ISO 14040 (2006) Environmental management: life cycle assessment: principles and framework

Jasiulewicz-Kaczmarek M (2011) Maintenance performance: overall equipment effectiveness. In: Lewandowski J et al (eds) Improvement of production processes, a series of monographs, Lodz, pp 71–91

Jasiulewicz-Kaczmarek M (2012) Socio-technical integrity in maintenance activities. In: Vink P (eds) Advances in social and organizational factors, CRC Press, pp 582–592

Jasiulewicz-Kaczmarek M (2013) Sustainability: orientation in maintenance management: case study. In: Golinska P et al (eds) Eco-production and logistics. Emerging trends and business practices. Springer, Heidelberg, pp 135–153

Johansson G, Winroth M (2010) Introducing environmental concern in manufacturing strategies: implication for the decision criteria. Manage Res Rev 33(9):877–897

Kaplan RS, Norton DP (1992) The balanced scorecard: measures that drive performance. Harvard Business Review

Komonen K (2002) A cost model of industrial maintenance for profitability analysis and benchmarking. Int J Prod Econ 79(1):15–31

Lawrie G, Cobbold I (2004) Third-generation balanced scorecard: evolution of an effective strategic control tool. Int J Prod Performance Manage 53(7):611–623

Loska A (2012) Remarks about modelling of maintenance processes with the use of scenario techniques. Eksploatacja i Niezawodność: Maintenance Reliab 14(2):92–98

Mikołajczak P (2011) Prediction of changes in the technical condition using discriminant analysis. Diagnostyka. Diagn Struct Health Monit 4(60):13–20

Mokashi AJ, Wang J, Vermar AK (2002) A Study Reliability-Centred Maintenance Marit Oper Mar Policy 26:325–335

Moubray J (1995) Reliability-centred Maintenance. Butterworth-Heinemann, Oxford

Nakajima S (1988) Introduction to TPM. Productivity Press, Cambridge

Napiórkowski J, Szczyglak P (2011) Analysis of factors in the decision process of purchasing farm tractors. Inżynieria Rolnicza 9(134):145–152

Napiórkowski J, Mikołajczak P, Wójcicki J (2011) The influence of power transmission poles located on the field on operational indicators of machine units. Inżynieria Rolnicza 5(130):199–205

Narayan V (2012) Business performance and maintenance. How are safety, quality, reliability, productivity and maintenance related? J Qual Maintenance Eng 18(2):183–195

Niu G, Yang B-S, Pecht M (2010) Development of an optimized condition-based maintenance system by data fusion and reliability-centered maintenance. Reliab Eng Syst Saf 95(7):786–796

Parida A, Chattopadhyay G (2007) Development of a multi-criteria hierarchical framework for maintenance performance measurement (MPM). J Qual Maintenance Eng 13(3):241–258

Pawlewski P, Borucki J (2011) "Green" possibilities of simulation software for production and logistics: a survey. In: Golińska P, Fertsch M, Marx Gomez J (eds) Information technologies in environmental engineering. New trends and challenges. Springer, Heidelberg, pp 675–688

Pawlowski E, Pawlowski K, Wachowski M (2006) Wdrażanie systemu TPM w warunkach przedsiębiorstwa międzynarodowego. In: Trzcielinski S (ed) Zarzadzanie we współczesnym przedsiębiorstwie, Monografia, Politechnika Poznańska, Poznań, pp 31–46

PBS DGA SA (2012) Biznes a ekologia: postawy wobec ochrony środowiska, http://www.euractiv.pl/wersja-do-druku/analizy/biznes-a-ekologia-postawy-wobec-ochrony-rodowiska-003228. Accessed 8 June 2012

Pinjala SK, Pintelon L, Vereecke A (2006) An empirical investigation on the relationship between business and maintenance strategies. Int J Prod Econ 104:214–229

Quezad C, Cordova FM, Palominos P, Godoy K, Ross J (2009) Method for identifying strategic objectives in strategy maps. Int J Prod Econ 122:492–500

Rosqvist T, Laakso K, Reunanen M (2009) Value-driven maintenance planning for a production plant. Reliab Eng Syst Saf 94:97–110

Shetty D, Ali A, Chapdelaine JJ (2009) A model for the total manufacturing assessment and implementation. J Adv Manufact Syst 8(2):117–136

Szafrański B (2011) Rola środków smarnych w utrzymaniu ruchu, Inżynieria & Utrzymanie ruchu zakładów przemysłowych, 7/8

Takata S, Kimura F, van Houten FJAM, Westkämper E, Shpitalni M, Ceglarek D, Jay Lee J (2004) Maintenance: changing role in life cycle management. Ann CIRP 53(2):643–656

Thakkar J, Dashmukh SG, Gupta AD, Shankar R (2007) Development of a balanced scorecard: an integrated approach of interpretive structural modeling (ISM) and analytic network process (ANP). Int J Prod Perform Manage 56(1):25–59

Zingales FGG, O'Rourke A, Orsatto RJ (2002) Environment and socio-related balanced scorecard: exploration of critical issues, INSEAD centre for the management of environmental resources (CMER), www.insead.edu/facultyresearch/research/doc.cfm?did=1235. Accessed 3 June 2012

Żółtowski B (2008) Technical diagnostics of folded objects. Dir Dev Diagnostyka 3(47):101–110

Part III
Improving Eco-efficiency: Case Studies

Assessment of Bangkok Metro Accessibility for Developing Integrated Strategies Using Sustainable Indicators

Duangporn Prasertsubpakij and Vilas Nitivattananon

Abstract This chapter aims to systematically develop a set of sustainable indicators regarding multifaceted aspects for assessing accessibility performance of the metro systems towards integrated strategies based on BMR empirical case. This effort elucidates three-functional steps of indicator selection including pre-selection of hypothesized indices based on literature reviews, application of the Weighted Index method via multi-stakeholder engagement, and formulation of the indicator set with the factor analysis and reliability test. A multi-dimensional accessibility assessment model was divulged as the major outcomes in which the potential effects of psychosocial and socioeconomic segments influence service accessibility in the study area. Discussions on sustainable indicators balancing different dimensions of accessibility evaluation and integrated strategies would contribute to accessibility-based knowledge and potential propensity to use the public transits towards transport sustainability.

Keywords Accessibility · Integrated strategy · Multi-dimensional indicators · Socioeconomics · Sustainability

---

D. Prasertsubpakij (✉)
Department of Applied Science, Chandrakasem Rajabhat University, Bangkok, Thailand
e-mail: honney.pra@gmail.com

V. Nitivattananon
Urban Environmental Program, SERD, Asian Institute of Technology (AIT), Bangkok, Thailand
e-mail: vilasn@ait.asia

M. G. Erechtchoukova et al. (eds.), *Sustainability Appraisal: Quantitative Methods and Mathematical Techniques for Environmental Performance Evaluation*, EcoProduction, DOI: 10.1007/978-3-642-32081-1_8,

1 Introduction

Comprehensive assessment and indicator selection process seem to exclusively affect the efficient environmental planning towards the city's desired sustainability goals. To provide decision-making on infrastructure and services development projects, city planners require baseline data and trends, evaluation performance on particular development issues and jurisdictions, and alternative solutions for successful planning. When considering public transportation accessibility plan as an intervention for enhancing the propensity to use public transits in metropolitan city, effective assessment mechanisms are vital to acquire the understanding of existing urban transportation system performance and the provisions of effective accessibility options based on different criteria. As evidence of Bangkok Metropolitan Region (BMR)[1] where private cars continuingly dominate the city transport vis-à-vis other modes as of the number of vehicle-kilometer traveled (VKT) (Perera 2006) exacerbating congestion and pollution, the improvement of public transit accessibility would be expected to be redesign for generating mode shift from car to rail-based transportation and provide sustainable choices on BMR future transport. In BMR case, the most contentious debates about continuing car dependent phenomenon in the city involves the difficulty of people in access to mass rapid transit systems (e.g., rapid elevated and underground transport) brought by the ineffective solutions based on traditional ways of metro accessibility development for instance, providing fare reduction or improving metro access facilities without concerning the needs of different user groups. It also relates to conventional strategies which have inherent limitation of holistic accessibility assessment as a great gap. The major scope of the study is to provide appropriate sustainable indicator selection process that can be applied for evaluating not only metro accessibility performance, but also other organization's environmental performance in cities.

1.1 Definitions of Sustainable Indicators

Conceptualization of sustainable indicators emphasizes the integrated consideration of three aspects of economic, social and environment in performance evaluation of development projects as measurable and desirable ways to achieve long-term sustainability goals. Sustainability enumerates one of the most fundamental human desires to bring about a better future world. It was exhibited as an agenda to simultaneously tackle the global environmental problems and to enhance the economic development of the poor (Newman and Kenworthy 1999). The definition of

[1] Bangkok Metropolitan Region (BMR) is called as Greater Bangkok comprising Bangkok (Thailand's capital city) plus five contiguous provinces namely Pathum Thani, Nonthaburi, Nakhon Pathom, Samut Prakan and Samut Sakhon.

sustainable development (SD) cited from Brundtland Report addressed as the development which delicate balance amongst the human needs of the present and future generations (World Commission on Environment and Development 1989). This focuses on fairness between intrageneration and intergeneration that often referred to as "the three-legged stool" of sustainability (May and Crass 2007). The most popular approach to gauging SD has been the employment of indicators and indices, and index being and an amalgam of more than one indicator (Liverman et al. 1988 and Quarrie 1992 cited in Bell and Morse 2003). Terminology is rather fluid, with labels such as sustainability indicators and indicators for SD being employed in various contexts. An 'indicator' may be defined as 'an operational representation of an attribute (quality, characteristic, property) of a system' (Gallopin 1997 cited in Bell and Morse 2003).

Regarding sustainable transportation indicators, there is a great concern to capture the balance set of indicators according to three sustainable aspects to measure a transportation performance in order to ensure transportation sustainability. To comprehend a sustainable transport planning, it takes a great deal of selecting an appropriate set of indicators that insists sustainability aspects (Litman 2007). Transport indicators can judge whether the transportation system and its impacts are good or bad depending on the community it serves.

In this chapter, the term "sustainable accessibility indicators" is developed that has been mentioned in many literatures (le Clercq and Bertolini 2003; Cheng et al. 2007; Curtis 2008). Developing sustainable accessibility indicators can be used as a key tool to design sustainable transport policies and plans by analyzing problems and evaluating alternative solutions for transportation sector, by which three key areas of sustainable development including economic, environment and social were considered. Although the words of transport accessibility and sustainable development are currently very different, a marriage between two terms could be magnificent (Gladwin et al. 2004). Sustainable transport accessibility involves broadly enriched the theoretical understanding of accessibility characteristics. It should be more concerned about social, economic, land use, temporal, and behavioral components (Cheng et al. 2007) besides transport components (travel time, cost and effort) (Geurs and van Wee 2004).

1.2 Metro Accessibility System: Multi-Components and Interactions

To develop accessibility assessment framework, this approach possess the combination of various components of accessibility and its interactions as illustrated in Fig. 1. The study also merges a perspective of metro accessibility and sustainable development by concerning more social, environment and economic in the way of influencing people to access metro services that leads to increase mobility needs, productivity, environment, and quality of life that creates sustainable transportation.

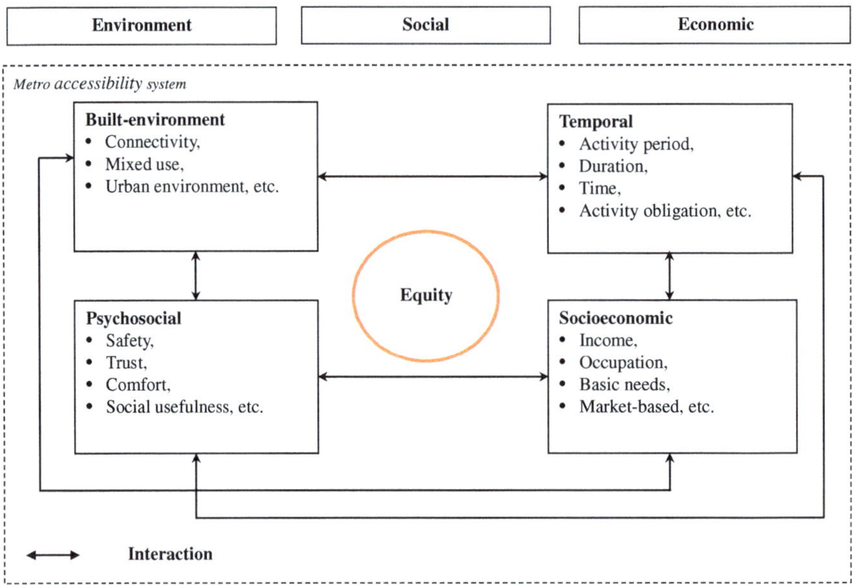

Fig. 1 Multi-dimensional accessibility components and interactions

This concept ravels the gap of a holistic accessibility study by formulating the balanced integration of multifaceted concepts of accessibility, that is, built-environment, socioeconomic, psychosocial, temporal, and equity aspects that compatible with sustainability criteria referred as the term a sustainable metro accessibility, which provides practical practices to pursue a path of sustainable transportation. This innovative approach demonstrates the interaction of the accessibility characteristics between environment, people (individual), time, and activities within metro accessibility assessment systems that parallel to the meaning of sustainable access or true access. It is important by this approach that accessibility assessment should concern not only typical user groups but also disadvantaged groups, such as women, the elderly and the disabled, to enhance equitable access for all user groups.

1.3 Metro Service Situation in Bangkok Metropolitan Region, Thailand

With rapid urbanization, the population of Bangkok Metropolitan Region (BMR), Thailand is over 10 million with severe peak hour traffic speeds about 10 km per hour throughout much of the metropolitan region. The authorities have a big effort to attract investors to develop metro systems in the late 1980s and 1990s, consisted of three lines- Blue, Red and Green. During the period, the Red and Blue Lines were planned and designed as elevated, called BTS skytrain systems. In 1994, the government made a decision to require the MRT lines to be built subway or mass rapid transit (MRT) within a 25 square km zone in inner-city Bangkok. The decision resulted in that Bangkok Land Public Company Limited withdrew from the Blue Line project, which subsequently became an MRTA project. Underground construction started in the year 1997 on the USD 3.08 billion. MRTA granted the 25-year concession to Bangkok Metro Company Limited (BMCL) in the year 2000 to operate the MRT lines which officially opened for services in 2004 (Phang 2007). The 15 km skytrain (green line) was built to serve downtown Bangkok and its alignment included two main routes, namely Si Lom and Sukumvit routes. In the year 1991, Tanayong Public Company Limited, a Thai real estate company, was selected from BMA to be purely privately financed because the company proposed the reasonable fare. And then, Tanayong developed a separate company, namely the Bangkok Transit System Company (BTSC), to build and own the concession. The concession contract provided for fare in-creases every 18 months in consistent with the increase in consumer price index. In the opening year, number of BTS passengers has been below the forecasted 570,000 level (just one-quarter forecast). The passengers in the year 2010 carried an average of 350,000 passengers/day.

Regarding organizational arrangement, there are various agencies with responsibilities in the metro sector in BMR, as shown in Table 1. These agencies involve BMA, transport department, public enterprises (MRTA) and metro companies (BMCL and BTSC). Moreover, there are national governments, provincial governments (authorities of peri-urban Bangkok) and other authorities that are

Table 1 The existing agency functions of metro services in BMR, Thailand

Existing functions related to metro accessibility	BTS skytrain	MRT subway
Policy and planning		
Line extensions (connectivity improving)	OTP	OTP
Land use and metro accessibility	Department of city planning BMA	Department of city planning BMA
Program development and management of infrastructure provision		
Design	BMA	OTP, MRTA
Facility construction	BMA	MRTA
Delivery of works	BMA	MRTA
Maintenance facilities	BTSC	BMCL
Service delivery on operation and maintenance		
Service delivery provisions	BTSC	BMCL
Ticketing and marketing	BTSC	BMCL
Service specification	BMA	MRTA
Contracting and contract compliance	BMA	MRTA

BMA Bangkok Metropolitan Administration, *BMCL* Bangkok Metro Company Limited, *BTSC* Bangkok Mass Transit System Public Company Limited, *MRTA* Mass Rapid Transit Authority, *OTP* Office of Transport and Traffic Policy and Planning.

responsible to implement the metro services. It is unavoidable that a large number of agencies related to metro planning, operation and implementation creates conflicts due to uncoordinated works. It is always a difficult job in the pattern of multi-agent planning and implementing form.

Bangkok's planning bodies have come up with two types of metro systems for Bangkok and peri-urban Bangkok. The administrative impact of a cooperative urban development in BMR can be viewed from no integrated fare and ticketing systems between BTS and MRT system. With this problem, it is inconvenient to use as an access barrier imposed by 'standalone' concessioning (World Bank 2007). It might be relatively difficult to put an integrated ticketing system into practice due to each public transit operator supplying its own equipment. This concern is an example of metro access problems relating to institutional aspect in BMR.

The trends of metro service improvement according to this respect should be considered in integrated metro systems (both BTS and MRT system) with bus and paratransit systems as major feeder modes in terms of cooperation organization and fare policy. Currently, people in BMR using various feeder options access metro services. It is unfortunate that there had been little formal integration of other public transports in BMR. It should be considerable the opportunity to restructure bus and van routes to better serve metro lines in the catchment area.

1.4 Access Problems on Metro Services Regarding Multi-Aspects

According to built-environment aspect, the majority of people in BMR live far from stations (Prasertsubpakij and Nitivattananon 2012) and metro lines lack of a total coverage of the network around and through the large settlements and residential areas outside of the inner Bangkok. Therefore, metro systems were not built to transport passenger from their homes to their activities in the inner city or the Central Business District (CBD). To date, people in BMR would use the metro services, in case they can access to feeder service that could provide satisfactory service quality and reasonable price for the whole trip of metro travelling. Access to stations always involves a mixed-access mode pattern for example, at Saphan Taksin Station (the station is located near the Chao Phraya River), people use multi-modes i.e., walk + bus + boat, walk + van + boat, hired-motorcycle + - bus + boat (except for elderly and handicaps). For the other stations, the access pattern is walk + paratransit, walk + bus, walk + bus + paratransit, bicycling or car only. This occurrence has an effect on travel costs, time and satisfactions across user groups. In addition, the individual costs from the one way trip from origin to destination reach 100 THB.[2] In this case, users who travel 10 trips per week (daily

[2] *THB* Thai baht (1 THB = 0.0326 US dollars, as of November 2012).

use) have to pay 4,000 THB for travel costs per month. If people who earn less than 10,000 THB/month, they need to spend half their salary for travel. However, if the estimated costs from travel are very high, they have tried to reduce their trips, found incentive and used mixed-modes instead.

In terms of temporal characteristics, time allocation and duration of activities or time restriction can affect the accessibility that is less concerned on metro planning and operating. In the morning peaks, the origin of metro trip comes from respondents' house location where far from metro stations. People travel by variety of feeders from their place to access closet station. Most of passengers found in this period are daily-users. The majority of trip purpose involve mandatory trip, such as go to work or go to school. For evening peaks, in contrast, people have short access distance from their activities (school or work place) to station. However, the destination of trip in this period has various possible alternatives, such as go back to their home or go for social purposes. In case of going home, they have to catch variety of modes to reach their houses. In during 10 am–16 pm, metro passengers access metro for shopping, having lunch or recreational purposes. There are various user groups found in this period in both of work and non-work groups. This includes disadvantaged groups, such as non-work women, the elderly and disabled people. In this period, feeder modes are scarce and irregular. People who travel in off-peak hours always face transport connectivity problems and inconvenient mobility.

According to access behavior of different user groups, researchers traced 25 passengers each station for investigating access and egress behavior of different user groups. The results from Table 2 show that access to and egress from station creates problems to both men and women in waiting and catching feeder modes. At night, women are vulnerable in park and ride that were located to walk distance from station. Women also have to cope with overcrowded and irregular public transport in access to and egress from station. Disabled people continue to face problems related to access mode opportunities, negative attitudes and environmental barriers. Women with disabilities face certain unique disadvantages compared with disabled men in many cases, e.g., feeder mode availability, special facilities for the disabled, universal signaling, elevators and sidewalk. The elderly people are obviously disadvantaged at present as to their personal accessibility. These groups need special attention when access-enhancing policies are designed. Women and disabilities in old edge suffer economic disadvantages due to gender biases in labor markets, person coverage and income generation opportunities. Despite, they got incentive from metro service but they have to spend more to catch taxis because of lack of convenience in catching bus or other modes.

2 Literature Review

Various previous researches applied some indicators to measure transport accessibility performance; albeit they interpreted the term "access" or "accessibility" in only a single-view of the sense of physical accessibility (Zegras and Srinivasan 2007;

Table 2 Access and egress behaviour of users at Bangkok metro stations

Identifiable groups	Access/egress modes	How to catch modes	Access/egress cost	Walking access	Access facilities	How they get service information	Incentive availability	Queuing and ticketing	Safety environment
Men	Use mixed modes	Manually	Vary by access mode types	They can reach to service within 5 min walk	Some of them complained about inconvenience of facilities	Manually and spend short time to access information	If they buy smart card, they will get discount	Only users who use single ticket and spend more than 5 min in case of rush hours	Few cases of accidents at station
Women	Use mixed modes except motorbikes (rarely)	Manually	Vary by access mode types	They can reach to service within 5 min walk	Women feel uncomfortable with sidewalk facilities	Manually and spend short time to access information	If they buy smart card, they will get discount	Only users who use single ticket and spend more than 5 min in case of rush hours	Women faced uncomfortable environment if station was crowded Women tried to avoid to go after 10-midnight because feeders were not allowed and unsafe

(continued)

Table 2 (continued)

Identifiable groups	Access/egress modes	How to catch modes	Access/egress cost	Walking access	Access facilities	How they get service information	Incentive availability	Queuing and ticketing	Safety environment
Elderly	Always use paratransits and cars	Manually/go with family/call taxi services	More than 50 % spend > 50 THB[a] for access costs due to taxi prices/gasoline prices	The majority spend time about 10 min walk	Most of them face difficulties with sidewalk/going by staircases	Most of them face difficulties to access information by themselves and always come to ask staff	Yes	No	Accidents can be occurred when they use staircases/escalators/crowded situations
Disabled	Limited only paratransits and cars	Manually/go with family/call taxi services	More than 50 % spend > 50 THB[a] for access costs due to taxi prices/gasoline prices	The blinds/visual impairments can reach to service within 5–10 min walk More than 50 % reach to service by getting some help from staff	The blinds always face problems with signaling	Get information from staff	If they show the disabled ID card, they will get free trip	No	They feel unsafe due to unavailability of the special access facilities

[a] *THB* Thai baht (1 THB = 0.0326 US dollars, as of November 2012)

Odoki et al. 2001; Zhu and Liu 2004) through how to find the effective way to help people to reach their destination or how to facilitate reaching (Tyler 2002). The metro accessibility in a view of physical perspective mainly indicates how it is efficient in terms of transport characteristics that describe the transport system, expressed as the disutility for an individual to cover the distance between an origin and a destination including the amount of time for traveling, waiting and parking, costs, effort, transit system's location and its characteristics. In this case, there are various criteria and indicators for measuring physical accessibility. Many efforts are found in previous researches e.g., Wardman et al. 2007; O'Sullivan et al. 2000; Guan et al. 2007; Guo and Wilson 2007; Chalermpong and Wi-bowo 2007; Vandenbuloke et al. 2009; Chalermpong 2007; Rodriguez and Targa 2004; Bollinger and Ihlanfeldt 1997; Bowes and Ihlanfeldt 2001; Odoki et al. 2001.

For example, Zhu and Liu (2004) measured the impact of the new mass rapid transit (MRT) lines on accessibility in Singapore by integrating GIS technique. The study identified the assumptions related to average speed of MRT, time transferring, waiting time during peak hours, bus feeder services and the shortest street paths from origin to end destination.

Accessibility assessment framework of Zhu and Liu (2004) was conducted by using the network time matrix function by classifying accessibility into four purposes including accessibility to CBD, to commercial and industrial activities and to working population with illustrating by before and after scenario. The findings found the spatial variations of accessibility in relation to working population, industrial and commercial opportunities. Another is O'Sullivan et al. (2000) developed desktop GIS application for accessibility assessment. The study was taken by the knowledge ground of the space–time accessibility measurement. The generation of isochrones (lines of equal travel time) for journey by public transport in Glasgow such as bus and underground services was investigated. The accessibility study in the trip include an initial walk to access a public transport stop, a number of interchanges between services of the same or different modes, and a final walk. However, the summary of literature review related to accessibility assessment study is classified into two groups including the assessment of metro accessibility performance and the accessibility assessment regarding transport components as presented in Table 3.

Nevertheless, the numerous definitions of accessibility from previous studies are still far from contributing meaningfully to the goal of true access or sustainable accessibility. Sustainable access on transportation should go beyond traditional practice of accessibility assumptions in perceiving holistic views of accessibility systems. This concept focuses on reaching to a destination and satisfying a human need, compared to only having the potential to access the resource with regard to the relationship of land use systems, transport networks and mode availability, travel behavior, temporal conditions, and costs (Becker 2004).

Table 3 Summary of metro accessibility measurement studies

Metro accessibility studies	Purposes	Indicators	Author (year)
The assessment of metro accessibility performance	To measure the impacts of metro accessibility related to property values	Walking time Walking distance Straight line distance	Chalermpong (2007); Bae et al. (2003); Rodriguez and Targa (2004); Bollinger and Ihlanfeldt (1997); Bowes and Ihlanfeldt (2001)
	To measure the impacts of metro accessibility related to spatial variation or location pattern	The potential measure of accessibility (based on the attraction of destination/distance, the distance decay parameter, the number of origins/destinations Gravity model to quantify accessibility based on location, distance, travel time, and job opportunity	Zhu and Liu (2004); Vandenbuloke et al. (2009); Cheng et al. (2007); Townsend and Zacharias (2010)
The accessibility assessment according to transport components (e.g., travel time, travel cost and distance)	To measure accessibility related to efficiency improvement	Access and egress time Aggregated accessibility measures and the space–time/geographical accessibility framework	Wardman et al. (2007); O'Sullivan et al. (2000)

Source as cited in the table

3 Sustainable Metro Accessibility Indicators

This section involves how to select sustainable metro accessibility indicators for metro accessibility assessment framework. The main steps of indicator selection are based on participatory approach and intensive information-gathering from key informants interview, questionnaire survey and secondary data collection. Weighting process was used to select sustainable metro accessibility indicators according to five-point Likert scaling regarding the rate of linkage with sustainable metro accessibility through 30 experts from multi-disciplinary of related fields.

3.1 Pre-selection of Sustainable Indicators: Literature Review

In the first step, a given set of indicators was selected from literature review and key informant interview. A tentative set of metro accessibility indicators was developed by integrating various aspects of accessibility with specific definitions. A set of hypothesized indicators was categorized into sustainability criteria including social, environment and economic groups. Only some of them have been found to be sustainability indicators, and these are the ones that should be given foremost consideration in order to applied for integrated accessibility assessment approach. Moreover, the selection proves was considered underlying the role of equity and justice.

3.2 Indicator Selection Process: Stakeholder Involvement

Next step is related to participatory approach. 30 representatives of various groups of stakeholders, such as local authorities (BMA), metro business [Bangkok Mass Transit System Public Company Limited (BTSC) and Bangkok Metro Public Company Limited (BMCL)], Mass Rapid Transit Authority (MRTA), Office of Transport Policy and Planning (OTP) and other interested groups were chosen and taken into account for indicator selection. In this stage, questionnaires comprised of the indicator set were sent by e-mail and by post to interdisciplinary experts related to multi-dimensional accessibility in order to weigh a score using a five-point Likert item (Likert 1932) and give us a suggestion. All experts were asked to rate the importance of each metro accessibility criteria, sub-criteria, and indicators to the sustainable metro accessibility according to a five-point Likert scale ("1" = an accessibility indicator reflecting the lowest relevance to the issue of sustainability while "5" = an accessibility indicator reflecting the most direct relevance to the issue of sustainability). The study applied average weighted index to select favorable indicators (based on 70 % up).

$$\textit{Weighted Average Index (WAI)} = (\mathrm{fss}(1.0) + \mathrm{fs}(0.75) + \mathrm{fps}(0.50) + \mathrm{fns}(0.25))/(\mathrm{ni} * \mathrm{N}) \quad (1)$$

where

fss frequency of strongly satisfied;
fs frequency of satisfied;
fps frequency of partially satisfied;
fns frequency of not satisfied;
ni number of item;
N total number of observations.

At this stage, a total of 22 indicators and criteria are left as shown in Table 4 and Fig. 2. Only some of them have been found to be sustainability indicators, and these are the ones that should be given foremost consideration in order to be applied for integrated accessibility assessment approach.

3.3 A Set of Sustainable Metro Accessibility Indicators

Albeit a set of indicators were purposely developed for metro accessibility assessment framework, this indicator selection process can be vindicated and then applied for opting sustainable indices for different types of environmental performance evaluation. According to weighting process, all criteria were selected with score >70 %. Figure 2 illustrates the results of indicator selection. The findings demonstrate that three sub-criteria are rejected from weighting process such as weather, socializing and attitudes because of low linkage with sustainability concern. It is automatically ignored seven indicators belonging to such sub-criterion. Twenty two indicators (Table 4) were accepted by experts with the rationale of highest, high linkage with sustainable metro accessibility based on various reasons. For example, park and ride indicator (0.682) was rejected by weighting process since the experts gave opinions that such indicator supports car usage leading to unsustainable urban transportation. Regarding social equity criterion, indicators in relation to opportunity to access of different user groups (0.727) and equality to access of different user groups were strongly accepted by the process based on the reason of long-term social sustainability and equity perspective. When considering temporal aspect, time and trip for out-of-home activities (0.636 and 0.614, respectively), and situational factors (0.614) were disregarded by reason of the characteristics of changeability. With economic consideration, only spending capacity (0.682) and perceived quality (0.5) indicators were ignored by with the same reasons.

Table 4 A tentative set of sustainable metro accessibility indicators

Aspect	Indicators	Detailed indicators
Built-environment	1. Land use mix and diversity of uses	Land use mix diversity indicators
	2. Potential of connectivity	Degree that roads and paths are connected and allow direct travel between destinations
	3. Compact and centeredness	Percentage of residential, commercial, employment and other activities nearby site stations within 500 m radius
	4. Attractiveness of metro utilities design	Level of attractiveness summations of utility design adopted to optimize the use of metro services
	5. Urban environment	Level of attitudes related spatial knowledge, modal experiences, paths and access behavior
Psychosocial	6. Information availability	Satisfaction level of continuous availability and usability of obtained information sources
	7. Attitude of continuity of route	Perception on transport and information provision, or any support for understanding metro route links and transferring
	8. Comfort	Satisfaction level of people on cleanliness, lighting, weather, ventilation, shade, etc
	9. Trust	The extent to which attitudes of willingness to confidence in ability, perceived control and quality of management
	10. Social usefulness	The extent to which attitudes on positive feelings of users on metro services on perceived support
	11. Safety	Sufficient level of commitment of safety planning and operation
Socioeconomic	12. Market-based	Attitude measurement on "do you get what you pay for"
	13. Basic needs	Level of attitudes on how metro services response their needs and desires
	14. Affordability by income	Personal income per month (Baht per month)
	15. Occupation	Occupation of respondents

(continued)

Table 4 (continued)

Aspect	Indicators	Detailed indicators
Equity	16. Opportunity	The extent to which attitudes of respondents think that they are taken into account in an equal manner
	17. Equality	The extent to which attitudes of respondents think that they receive equal benefit from services
Temporal	18. Activities obligation	The distance of activities location from stations (km)
	19. Time obtaining	Total time (mins)
	20. Trip purpose	The characteristics of trip purposes
	21. Trip duration	Duration preference of metro trip
	22. Activity period	Time period of target activities

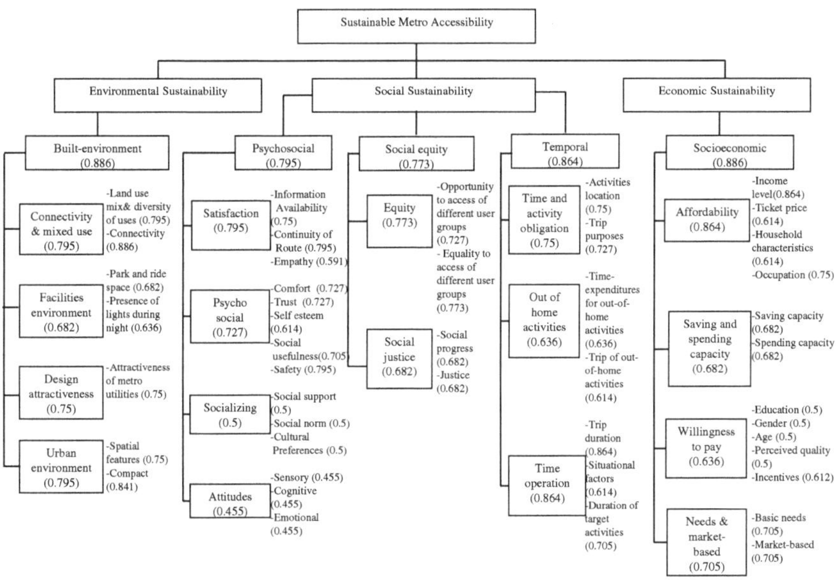

Fig. 2 Selected sustainable metro accessibility indicators by expert groups

4 Development of Metro Accessibility Assessment Model

4.1 Assessment Model Formulation

A formation of integrated metro accessibility assessment framework in order to assess accessibility performance of metro systems in BMR is based on the method of developing a sustainability assessment model of Yigitcanlar and Dur (2010). Factor analysis is applied to cluster twenty two indicators from above step. With a purpose of ensuring suitability for conducting factor analysis, the study applied Kaiser-Mayer-Olkin (KMO) test (Kaiser 1960) for measuring the adequacy of a sample in terms of the distribution of values for the execution of factor analysis and Barlett's test (Geourge and Mallery 1999) of sphericity for determining if the correlation matrix is an identity matrix (sig. 000), which would indicate that the model is inappropriate. The result of the KMO test is 0.603 and Barlett's test of sphericity illustrates a probability values of 0.000 (Table 5). Both tests indicate the suitability of the indicator set for factor analysis.

In the next step, factor extraction is carried out through factor loadings to find the correlations between each indicator. High factor loadings mean that the selected indicators are critical (Kline 1996). To determine the number of indicators retained, both eigenvalues and communality techniques are applied for the analysis. The results of total variance explained from SPSS show the eigenvalues of 10 indicators over 1 that should be retained (Kaiser 1960).

For the communality, referred as a proportion of variance of variable in common with each pattern, indicators with factor loading value less than are eliminated (see Table 6). Furthermore, the study also makes use of 'Scree test' (Cattell 1978) to consider number of indicators retained. The results of scree plot from SPSS generate the influential indicators that should be kept for assessment, which are the ones on the steep slope (Prasertsubpakij 2012). Table 6 shows ten indicators grouped by factor analysis approach. The researcher gave a new name for each acceptable indicator that covers the domain of the predictors being measured based on literature review and the opinions of experts.

4.2 Model Application and Discussions

One example of model application was used for assessing metro accessibility performance in BMR to scrutinize how user groups access metro services based on BMR empirical case Based on this study, 600 individual passengers at various stations were asked to rate the questionnaire that simultaneously considers accessibility aspects of spatial, feeder connectivity, temporal, comfort/safety, psychosocial and other dimensions. The model called as Users Disaggregated

Table 5 The results of KMO and Bartlett's test for indicating the suitability of the indicator set for factor analysis

Kaiser–Meyer–Olkin measure of sampling adequacy.		0.603
Bartlett's Test of Sphericity	Approx. Chi Square	8.483 E3
	df	351
	Sig.	0.000

Table 6 Integrated indicators selected by factor analysis and reliability test

Aspect	Indicator interpretation	Predictors included in the indicator	Reliability test[a]	Factor loading[b]	Item-total correlation[c]
Built-environment	Connectivity and mixed use (y5)	Potential connectivity	0.747	0.845	0.596
		Mixed use		0.776	0.596
	Urban environment (y9)	Compact	0.732	0.776	0.609
		Urban environment		0.517	0.609
	Design attractiveness (y6)	Attractiveness of metro utilities	–	0.710	–
Psychosocial	Psychosocial (y1)	Safety	0.865	0.896	0.857
		Social usefulness		0.896	0.857
		Trust		0.757	0.704
		Comfort		0.604	0.466
	Satisfaction (omitted)	Satisfaction of continuity of route	0.544	0.708	0.374
		Information availability		0.629	0.374
Socioeconomic	Affordability (y3)	Income	0.723	0.749	0.570
		Occupation		0.668	0.570
	Basic needs and market-based (y4)	Basic needs	0.848	0.835	0.736
		Market-based		0.834	0.736
Temporal	Temporal (y2)	Activity period	0.892	0.893	0.874
		Trip purpose		0.816	0.761
		Duration		0.752	0.734
	Time and activity obligation (y8)	Time obtaining	0.704	0.693	0.573
		Activity Location		0.433	0.573
Equity	Equity (y7)	Equality	0.717	0.779	0.559
		Opportunity		0.724	0.559

Note: [a] Internal consistency reliability: acceptable minimum level 0.7 based on Cronbach's alpha (Cronbach 1951; Nunnally1978)
[b] The amount of variance in each variable that is accepted for
[c] Items-total correlation indicates to what degree items into correlate with each other. Items with total item correlation of ≤ 0.4 are omitted

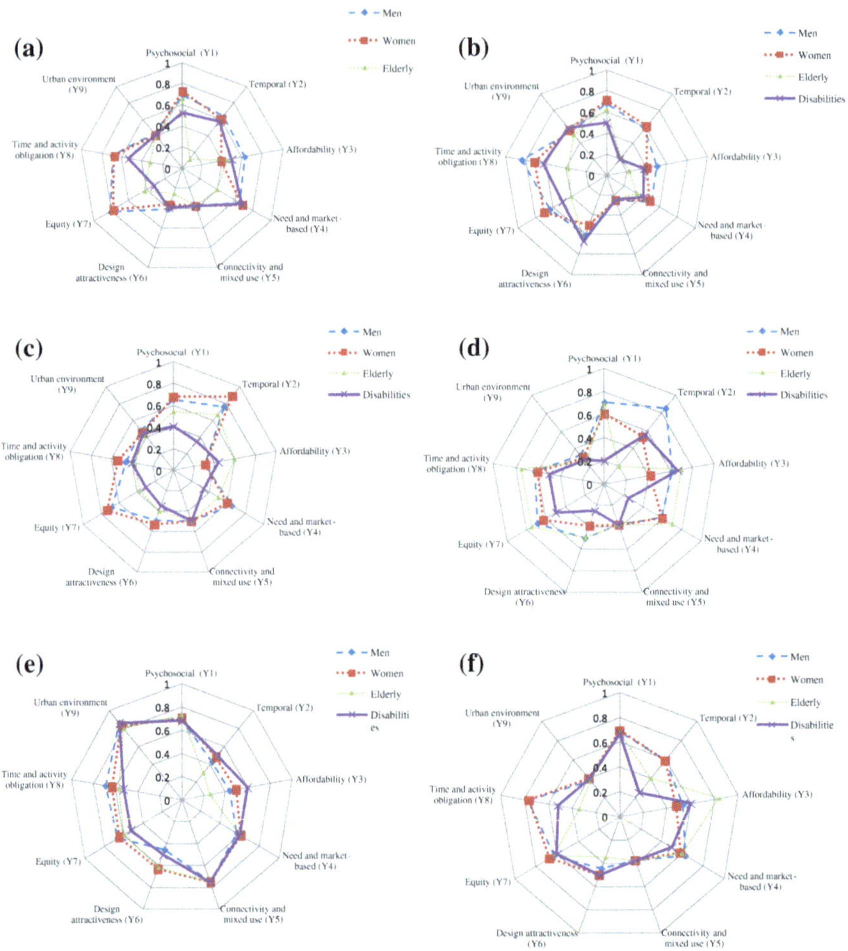

Fig. 3 Metro accessibility performances of selected stations disaggregated by user groups. *Note* **a** Accessibility scores 0.00 indicates very poor accessibility performance, ranging to accessibility scores 1.00 indicates very good accessibility performance. **b** *T*-test sig. at 95 % confidence level for male and female accessibility = 0.042 (t-values 2.638), for typical and disabled users = 0.027 (t-values 2.215), and for typical and elderly people = 0.363(t-values 0.910). *Source* Prasertsubpakij and Nitivattananon (2012)

Accessibility Model (see Fig. 3) elucidates the potential effects of psychosocial, comfort and safety and other dimensions to the disabled, elderly, in fact, not performing well in light of their needs and abilities on their experiences with accessibility. The model illustrates how socioeconomic segment influences more to the rights of women in services accessible. Figure 3 represents the multi-dimensional accessibility assessment for metro services across user groups of selected stations. The detailed examination shows that women, the elderly and disabled were found to have difficulty accessing services in several aspects.

There is various ways to assess metro accessibility performance. While, conventional assessment or one-side approach based on previous researches is popularly used as a common practice by planners in investigating the performance of metro access phenomenon that always attains unsustainable results, this assessment framework presents an alternative accessibility measurement unit, which can be useful for transport planners and decision-makers in local government, enterprise and metro companies to understand a true access perspective that usually is affected by many factors. Innovative evaluation framework are encapsulated the combination of versatile aspects of metro access regarding environment, social and economic aspects to ensure a sustainable metro accessibility. The model and data in this research are also being applied by consultants in those questions and assumptions which need to be appropriately adjusted. The integrated indicator model may be useful to measure the accessibility performance of across user groups.

4.3 Model Limitations and Recommendations

Though Users Disaggregated Accessibility Model captures the holistic view of accessibility consideration along with the balanced access priorities of different user groups, there are some limitations that come with the model as well. It lacks the comparative non-user based model for comparison purpose; hence non-user study is suggested to fulfill the gap particular in the behavioral and psychosocial responses, socioeconomic and barriers of non-users on using the services. The comparison of the access decisions to services amongst groups is recommended for areas of further research.

With our suggestion, the adjustment of the model or assessment framework could be involved in recognition of the measurement of preferences between traveler and non-travelers on mode choice decision underlying the behavioral economics and psychological approaches. Modification of model might consider the approach of geographical configuration and benchmarks. Regarding this point of view, user-based and non-user-based accessibility model with a focus on different conditions of urban zones across socioeconomic segments can be integrated. Another limitation is related to the narrow set of performance indicators that falls so far short of a complete analysis of economic and institutional impacts. For limitations on the assessment process, due to limited time and resources, the study selected only some stations for assessment framework. There is inevitably a balance to be struck between the sizes of the metropolitan areas to be modeled.

5 Multi-Dimensional Metro Accessibility Performance and Integrated Strategies Via Sustainable Indicators

5.1 Accessibility Performance of Bangkok Metro Services

This section aims to draw the illustration of various dimensions of accessibility performance based on BMR, Thailand empirical case by using the model developed by the study. The study was carried out as part of a field survey, by which a sample of metro users was selected from site stations of BTS skytrain and MRT subway in BMR, Thailand. The users were chosen by random sampling for interview. We asked them for e-mail address and phone number contacts for getting more information needed. 600 samples were reached in 3 months. Our questions are divided into three parts. First, respondents at stations answered the questions about individual trip characteristics. This includes residential location, trip purpose and destination. We recorded the information about gender, age, status, income, education, vehicle ownership and trip frequency. Second part of questionnaire assigned questions in relation to data supporting multi-dimensional accessibility assessment. The information can be used for synthesizing the current situations. The next part is a vital part. Participants were asked to fill questions according to influential factors. Opinions of access to metro services were discussed and suggested. Assessment results regarding multifaceted indicators of metro accessibility e.g., built-environment, psychosocial, socioeconomic, and temporal dimensions are more described as follows:

Regarding built-environment aspect, incompatibilities between transportation—land use—users—services as crucial concerns for metro accessibility restriction were found in the study based on our model. There is weak evidence of transport network connectivity in BMR. As non-efficient physical accessibility, it is manifestly seen that BMR land use illustrates sprawling configuration that affects people in difficult approachability from their houses to metro stations or from stations to their proposed destinations. Metro lines have not been well organized and poor coordinated with other public transport networks. BMA has a city planning vision in relation to improve public transport access for all but in reality; the implementation is very weak because of incoordination of various stakeholders and fragmented authorities in different geographical scales. Regarding the performance from built-environment indicator engaging integrated compact and urban environment parameters, the density of selected stations seem plausible to describe a spatial feature of sprawling growth. It results in low level of metro accessibility with mean values less than 0.5. The expansion of town can be defined as decentredness that supports car-oriented design and discourages the public transit usage. From the survey, sprawling increases travel time, for instance, the shortest average total time—less than 30 min—amongst the respondents are found only people who live nearby stations or within the neighborhood.

According to temporal aspect, activity engagement and time opportunity of different user groups accessing metro services are a key premise of accessibility performance. Its interaction associates the trip-making behavior of gender differences and other groups in social inclusiveness. Considering temporality (activity period, trip purpose, duration), the study demonstrates a good performance of metro accessibility with the mean values of 0.625 (Prasertsubpakij and Nitivattananon 2012). The rationale of this performance result is determined by the long period of mandatory trip (go to work or go to school), mainly in case of peak period. The peaks represent relatively high level of feeder choices, in which there are a large number of activities to be undertaken in full (Ashiru and Nielsen 2001). According to time and activity obligation, most of metro commuters try to save their time use on metro travel. The average total time use is calculated by the combination of access and egress time and time use on metro services. The findings illustrate that almost 60 % of respondents have the average total time <50 min. Compared with single-car or single-bus usage, metro travel provides efficient time savings. Regarding another component of this indicator, the performance is measured by the distance between end-station and activity-location (km) or the egress distance (km). It is not surprising that the majority (about 73 %) demonstrate a good performance with the lowest average egress distance <2.4 km (Prasertsubpakij 2012).

With socioeconomic consideration, there has been a great deal of debate over the issue of how to make metro services more affordable and accessible to all, especially in BMR. Ac-cording to the survey, about 70 % of respondents earn monthly income of 10,000–20,000 THB ($\approx$USD 285–270). People who earn less than 10,000 THB/month are counted around 17 %. In this case, most of them are unemployed and students. Meanwhile, average total expenditures per trip for individual is very high about 61–80 THB/trip, so passengers appear to be affordless. The mean value of level of metro accessibility regarding affordability aspect is 0.466 that indicates poor accessibility. The findings are supported by Ketraungroch (2008) that addressed metro transits in Bangkok can be of use for households whose monthly income is greater than 15,000 THB ($\approx$USD 435). Regarding basic needs and market-based indicators, the attitude test results attest that the majority of respondents feel partially satisfied about the basic needs and market-based indicator or feel that 'they get what they pay for' with 45.3 and 47 % respectively. However, people of each group share different opinions on this respect. In terms of how well the needs are met, all user groups need better side walks, especially in BTS stations, the specific instrumental facilities and feeder availability. However, the findings of overall performance regarding need and market-based indicator show high level of metro accessibility performance with mean values of about 0.6 (Prasertsubpakij 2012).

The performance based on psychosocial indicator including safety, social usefulness, trust, and comfort predictors represents the high level of metro accessibility with the mean values of about 0.691 (Prasertsubpakij and Nitivattananon 2012). It is apparent that the psychosocial effects that strongly associated with behavioural responses to access are found in a good perception. More than a half of respondents have positive feelings on perceived support or social support from media, family and education in order to provide them with easier access. In addition, the majority of metro commuters feel satisfied by indoor environment (e.g., cleanliness, lighting, weather, seat and comfort, etc.) For safety issue, 55.3 % of respondents feel confident about the safety policy, planning and management of metro systems as well as the reliability on services. There are 23.5 % of people who feel partially confident. Only 2.8 % of respondents have low confidence because they are afraid of crime opportunity and emergency situations. The overall performance regarding psychosocial respect seems to be a good perception; however, if we consider the equity access among different user groups. The performance is quite poor. Disadvantaged groups, mainly the elderly and disabled people express very low perception on this concern. They find it inconvenient to access the services and have faced difficulty due to indirectness of route and waste of time on transferring as well as creating opportunity costs.

5.2 Integrated Strategies

It was revealed by this study that BMR metro passengers faced unsustainable access that requires strategies and potential options to improve and to sustain metro services. Regarding the evaluation performance, it illustrated low level of accessibility performance in some aspects, such as affordability, equitable service quality and built-environment aspects. Hence, firstly, policy implications based on poor performance related to land use planning and path and facility design should be considered by city planners. Transit-Oriented Development strategy regarding mixed use development can play a key role. Secondly, innovative strategies should encourage and enable service usability for all. Improving station facilities through the development of single-fare system and appropriate physical infrastructures are needed. Thirdly, since psychosocial aspect, mainly behavioral responses to access of different user groups is a key concern for achieving equitable access for all. Gender differences in affordability and temporal aspects of metro accessibility can provide a special concern to improve key facilities, scheduling and fares for enhancing women's access to services. Another strategy should focus on strengthening formal and informal education and communication through multi-medias in order to foster a service confidence, to change cultural preferences from car usage to use public transit services as long-term access behavior change.

Another consideration based on the study is that context of behavior responses to access were discussed from individual actions, intentions and perceptions to metro accessibility as the absence of previous researches. It is defined as the social concern of metro accessibility systems that are relatively poorly understood. Social interactions within metro accessibility systems are important and powerful ways of illustrating individual access behavior that can sustain metro services. It presents a relationship between access and explanatory characteristics at a range of spatial accessibility patterns. Such social factors may affect how people with different abilities, ages, genders, and ethnicities access public transport and reflect access behavior and needs.

In addition, not only socioeconomic demographics but also personal access behavior and attitudes towards willingness to access metros are found to considerably affect the commuter mode choice that might discourage car preferences. Unsatisfied psychosocial indicators and difficulties of using metro services might show the negative effects on all public transport patronages. Negative attitudes towards comfort and convenience of metro accessibility might pose the potential of choosing car rather than metro services. Therefore, it is necessary for transportation planners to focus on the socioeconomic and psychosocial indicators towards the combination of other aspects of accessibility evaluation in order to enhance the performance of metro accessibility in cities.

6 Conclusions

The manifestation of the accessibility performance of Bangkok metro services explicates the significance of the balance integration of multi-dimensional aspects of accessibility appraising sustainability perspectives or the economic, social and environmental influences on the city plan. Our approach on combination of different components of accessibility is one example of shifting from one-sided approach popularly used as a common practice in investigating the transportation performance to cross-solutions that can be considered as a part of the sustainable appraisal process to ensure the decision of planners in order to achieve sustainable development (SD). This assessment framework presents an alternative accessibility measurement unit, which is valuable for transport planners and decision-makers in local government, enterprise and metro companies to understand a true access perspective that is usually affected by many factors. Currently, the BMR has ongoing projects that need to carefully design by addressing key needs of different user groups regarding multi-dimensional aspects. The model and data in this research are also being applied by consultants in those questions and assumptions which need to be appropriately adjusted. The integrated indicator model may be useful to measure the accessibility performance of other transit services or any related public transportation service across user groups. Our indicator selection approach can be applied for evaluation of sustainability of other organization's environmental performance as well.

References

Ashiru O, Nielsen S (2001) Temporal accessibility: a computational approach for assessing accessibility within rural and urban transport networks. In: Proceedings of the AET European transport conference, Cambridge

Becker U (2004) Access and accessibility, sustainable development, and transportation: how to measure efficiency in transport? In: Access to destinations—rethinking the transportation

future of our region. Proceedings of access to destinations conference, University of Minnesota, Minneapolis
Bell S, Morse S (2003) Measuring sustainability: learning by doing. Earthscan, London
Bollinger CR, Ihlanfeldt KR (1997) The impact of rapid rail transit on economic development: the case of Atlanta's MARTA. J Urb Econ 42:179–204
Bowes DR, Ihlanfeldt KR (2001) Identifying the impacts of rail transit stations on residential property values. J Urb Econ 50(1):1–25
Cattell RB (1978) The scientific use of factor analysis in behavioral and life sciences. Plenum, New York
Chalermpong S (2007) Rail transit and residential land use in developing countries: hedonic study of residential property prices in Bangkok, Thailand. Trans Res Rec 2038:111–119
Chalermpong S, Wibowo SS (2007) Transit station access trips and factors affecting propensity to walk to transit stations in Bangkok, Thailand. J East Asia Soc Trans Stud 7:1806–1819
Cheng J, Bertolini L, le Clercq F (2007) Measuring sustainable accessibility. Trans Res Rec 2017:16–25
Cronbach LJ (1951) Coefficient alpha and the internal structure of tests. Psychometrical 16:297–334
Curtis C (2008) Planning for sustainable accessibility: the implementation challenge. Trans Policy 15(2):104–112
Geourge D, Mallery P (1999) SPSS for windows step by step: a simple guide and reference. Allyn & Bacon, New York
Geurs KT, van Wee B (2004) Accessibility evaluation of land-use and transportation strategies: review and research directions. J Trans Geogr 12(2):127–140
Gladwin T, Kelbaugh D, Simon C, Sullivan J (2004) A complex systems approach to sustainable accessibility: the University of Michigan "SMART" project. In: Proceedings of access to destinations conference, University of Minnesota, Minneapolis
Guan H, Yin Y, Yan H, Han Y, Qin H (2007) Urban railway accessibility. Tsinghua Sci Technol 12(2):192–197
Guo Z, Wilson NHM (2007) Modeling effects of transit system transfers on travel behavior: case of commuter rail and subway in Downtown Boston, Massachusetts. Trans Res Rec 2017:11–20
Kaiser HF (1960) The application of electronic computers to factor analysis. Educ Psychol Meas 20:141–151
Ketraungroch P (2008) The effect of mass rapid rail transit on residential location pattern in Bangkok. Thammasat University, Thammasat
Kline P (1996) An easy guide to factor analysis. Routledge, Great Britain
le Clercq F, Bertolini L (2003) Achieving sustainable accessibility: an evaluation of policy measures in the Amsterdam area. Built Environ 29(1):36–47
Likert RA (1932) Techniques for the measurement of attributes. Arch Psychol 140:1–55
Litman T (2007) Developing indicators for comprehensive and sustainable transport planning. Trans Res Rec 2017:10–15
May T, Crass M (2007) Sustainability in transport: implications for policy makers. Trans Res Rec 2017:1–9
Newman P, Kenworthy J (1999) Sustainability and cities. ISLAND Press, Washington DC
Nunnally JC (1978) Psychometric theory (2nd edn). McGraw-Hill, New York
Odoki JB, Kerali HR, Santorini F (2001) An integrated model for quantifying accessibility-benefits in developing countries. Trans Res A 35(7):601–623
O'Sullivan D, Morrison A, Shearer J (2000) Using desktop GIS for the investigation of accessibility by public transport an Isochrone approach. Int J Geogr Inf Sci 14(1):85–104
Perera R (2006) Promoting travel demand reduction in transport sector in cities of Asian developing countries: Case of Bangkok. Workshop paper, Urban Environmental Management (UEM), SERD, Asian Institute of Technology
Phang SY (2007) Urban rail transit PPPs: survey and risk assessment of recent strategies. Trans Policy 14(3):214–231

Prasertsubpakij D (2012) Multi-dimensional accessibility assessment of metro systems for sustainable transportation in Bangkok Metropolitan Region. Dissertation, Urban Environmental Management (UEM), SERD, Asian Institute of Technology

Prasertsubpakij D, Nitivattananon V (2012) Evaluating accessibility to Bangkok Metro Systems using multi-dimensional criteria across user groups. IATSS Res 36(1):56–65

Rodriguez D, Targa F (2004) Value of accessibility to Bagota's Bus Rapid Transit Sys-tem. Trans Rev 24(5):587–610

Townsend C, Zacharias J (2010) Built environment and pedestrian behavior at rail rapid transit stations in Bangkok. Transportation 37(2):317–330

Tyler N (2002) Accessibility and the bus system: from concepts to practice. Thomas Telford, London

Vandenbuloke G, Steenberghen T, Thomas T (2009) Mapping accessibility in Belgium: a tool for land-use and transport planning? J Trans Geogr 17(1):39–53

Wardman M, Lythgoe W, Whelan G (2007) Rail passenger demand forecasting: cross-sectional models revisited. Res Trans Econ 20:119–152

World Bank (2007) Strategic urban transport policy directions for Bangkok. Report No. 41075. http://siteresources.worldbank.org/INTTHAILAND/Resources/333200-1177475763598/2007june-bkk-urban-transport.pdf. Accessed 12 August 2010

World Commission on Environment and Development (1989) Our common future. Oxford University Press, Oxford

Yigitcanlar T, Dur F (2010) Developing a sustainability assessment model: the sustainable infrastructure, land-use, environment and transport model. Sustainability 2:321–340

Zegras PC, Srinivasan S (2007) Household income, travel behavior, location, and accessibility: sketches from two different developing contexts. Trans Res Rec 2038:128–138

Zhu X, Liu S (2004) Analysis of the impact of the MRT system on accessibility in Singapore using an integrated GIS tool. J Trans Geogr 12:89–101

How District Energy Systems can be Used to Reduce Infrastructure Costs and Environmental Burdens

Terri Chu and Sandra Yee

Abstract District Energy Systems are increasing in popularity but their economic viability is usually analyzed narrowly in terms of capital costs and energy savings to the customer and revenues to the utility provider. Traditional feasibility analyses of DESs do not recognize the economic benefits to municipalities and regional governments. By having a DES, electricity demand during peak times can be reduced. A reduced demand means fewer electrical peaking stations will need to be built thereby saving regional governments substantial sums in infrastructure costs. Another noteworthy benefit is the cost of storm water retention. DES is an enabler of storm water retention technologies. By using DES rather than a traditional rooftop mechanical room, space is made for water retention technologies that could not otherwise be built. By reducing the amount of water flowing from a building site, municipalities reduce the risk of sewer overflows and can reduce the infrastructure required for storm water containment. Lastly, a DES produces thermal energy on a large scale and is technology neutral. The nature of DES allows for fuel diversity and flexibility. Should the cost of any one type of fuel increase dramatically in price, DESs have the ability to switch sources with minimal investment. DESs will allow municipalities to ensure their communities will be able to maintain reasonable fuel costs and a high standard of living. None of these economic benefits are included in current feasibility analyses yet they can be substantial. If these factors were included, the economic case for DES would be made quite easily and communities could then benefit from the reduced carbon

T. Chu (✉) · S. Yee
FVB Energy Inc, 3901 Highway 7, Suite 300, Vaughan, Ontario, Canada
e-mail: tchu@fvbenergy.com

S. Yee
e-mail: syee@fvbenergy.com

M. G. Erechtchoukova et al. (eds.), *Sustainability Appraisal: Quantitative Methods and Mathematical Techniques for Environmental Performance Evaluation*, EcoProduction, DOI: 10.1007/978-3-642-32081-1_9,

footprint for their heating and cooling. This makes a compelling case for municipalities to support DES in their communities or legislate building connections to DES.

Keywords District energy · Municipalities · Energy reduction

1 Addressing Energy Security and Reducing Infrastructure Costs in Power Generation and Storm Water Management

A District Energy System (DES) has the potential to help governments, at all levels, reduce operating and infrastructure costs. These opportunities, however, are often left unexplored. In a free market environment, DES development will only take place with a strong business case for a DES Utility and its customer base. By recognizing and crediting the cost savings that a DES affords governments, the business case for DESs would be strengthened and help policy makers to take advantage of the eco-efficiencies these systems provide.

Understandably, governments are averse to providing financial or human capital to support the private sector without a reasonable expectation of socio-economic benefits for taxpayers. While these capital intensive systems do not necessarily need government support to be economically viable, returns are often marginally below acceptable levels for investors. By helping the private sector breach the risk/reward threshold, municipalities could reduce their own operating costs by millions of dollars with minimal investment and risk.

To realize the additional benefits of DESs, it is crucial to understand what district energy systems are, the current business case for district energy, and the economic and environmental opportunities that are complementary to government objectives such as: power generation, storm water management, energy security, and economic growth.

2 District Energy 101

District energy is a technical solution for providing the thermal energy used for conditioning indoor spaces. DESs are generally comprised of three major components:

- A common or shared energy generating facility referred to as a central energy plant or community energy centre,
- A system of interconnected pipes that link the energy centre(s) to multiple buildings referred to as a distribution piping system or thermal grid,

- A thermal interface at the customer building referred to as a customer substation or energy transfer station.

In the case of district heating, hot water (or steam) is transported through a system of pipes and delivered to the customer buildings for space heating and domestic water heating. The heat energy carried in the fluid is extracted by the building's systems and the cooled water is returned to the central energy centre, in a closed loop piping system, where it is reheated for redistribution. Similarly with district cooling, chilled water is pumped through a network of pipes and the cooling energy is extracted by the building for air conditioning or process cooling and then the warmed water is returned to the plant to be cooled again.

Figure 1 illustrates that District Energy is not technology specific; multiple or single fuel sources and/or technologies can be employed including but not limited to: absorption chillers, ambient cooling, deep lake water cooling, combined heat and power, biomass incineration, waste incineration, geothermal, and conventional boiler and chiller technologies. Figure 2 illustrates how district energy typically works. Pipes run underground bringing heating and cooling energy to buildings from a central energy plant.

Economies of scale and advancements in technology have enabled DESs to achieve greater efficiencies than individual building systems. Serving multiple buildings from a common facility allows for large scale systems to be built that can accommodate state-of-the-art technologies in heating and cooling. For example, individual buildings could not cost effectively make use of natural lake water cooling or Combined Heat and Power (CHP). Using natural sources of cooling or

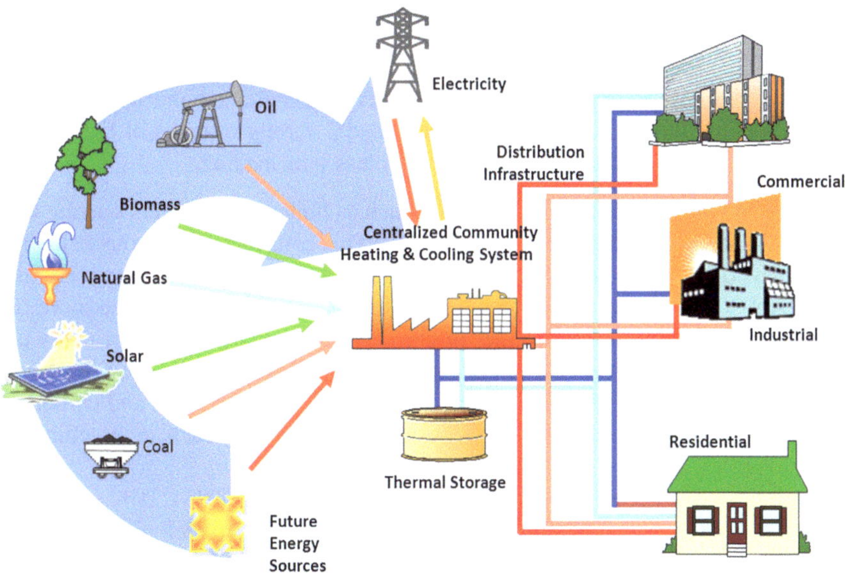

Fig. 1 DES can accept any form of energy source (IDEA 2012)

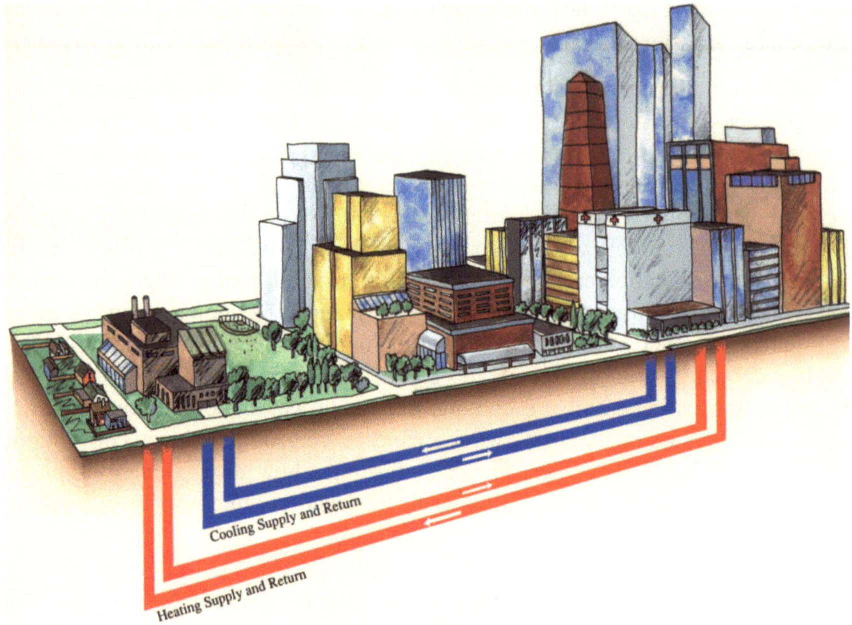

Fig. 2 Typical district energy system (IDEA 2012)

waste heat sources from manufacturing or electricity generation can reduce energy consumption for the production of thermal energy to near zero.

In addition, economies of scale in DESs enable the efficient use of conventional equipment by:

- Having purpose built facilities that are actively maintained and operated
- Aggregating thermal loads from multiple and a variety of types of buildings so that equipment can be run optimally with less part loading.

Thermal energy generating equipment such as boilers and chillers operate most efficiently at a single load factor (Chan 2002). This concept is analogous to the fuel efficiency of a car, reaching optimal fuel economy between 50 and 80 km/h (NRCAN 2012a, b). A car traveling outside of its 'sweet spot' will use more fuel per km travelled. Likewise, a boiler or chiller will use more energy for every unit of thermal heating or cooling when forced to operate above or below its 'sweet spot'. By having an aggregated load, a DES can 'stage' its boilers and chillers by operating only the number required by while maintaining each in its efficiency 'sweet spot'.

The earliest DES is dated as far back as the 1300s (CDEA 2011). The first known system distributed warm water through a series of wooden pipes in France. In 1877, the first commercial DES was built in New York (CDEA 2011). DESs are not a new concept. Though widely embraced in Europe, DESs have not seen the same rate of adoption in North America. Abundant and low cost energy supplies have reduced the urgency for conservation and innovation in thermal energy production. However, energy constraints, limited dollars, and environmental concerns are putting DESs on the public agenda. To address these concerns, policy makers can examine the synergistic opportunities that DES can provide.

3 The Current Business Case for DES

The current business case for DES fails to examine or value the economic and environmental benefit to other stakeholders such as industries, communities, and governments. The industry standard for determining the feasibility of investing in DESs is based solely on the economic benefit to both the building owner(s) (customers) and the DES utility.

For the DES Utility, the profitability of developing a DES is evaluated by comparing the initial capital investment costs to the expected cash flow over the life of the system. Capital costs include the cost to build the DES infrastructure: the energy centre, the distribution piping system, and the customer connections. Expenses are dictated by fuel, operating, maintenance, and administration costs. Revenue is based on a capacity charge to the customer as well as energy delivery charges. The price is dictated by the Business-As-Usual (BAU) cost to produce thermal energy for space heating, cooling, and domestic hot water. After a financial analysis, the DES Utility decides if the business case for DES passes the risk/reward threshold that the investor is willing to accept.

A similar economic analysis is performed by the building developer or owner comparing the price of the district energy service to the current (or estimated) capital and operating costs of providing building heating and cooling. The district energy service is priced competitively, equal to or below, the BAU model. The cost savings a building would realize by connecting to a DES is comprised of some or all of the following:

- Reduction in initial and/or replacement capital cost for major mechanical equipment including cost of associated space, electrical installation, and auxiliaries
- Fuel costs (i.e. natural gas for heating, and electricity for cooling)
- Cost of water, sewer, and water treatment
- Equipment operating and maintenance cost, including yearly preventative maintenance and ongoing repair/over haul costs
- Cost of labour, administration, and insurance
- Value of “freed up” roof space, greenhouse gas reduction, risk mitigation, and liability.

4 A "Broader" Business Case for DES

A broader business case for DES may be developed by understanding how DESs can create eco-efficiency opportunities for governments. Three such opportunities and their respective benefits can be found in:

- Reduction in electrical generating capacity,
- Storm water management, and
- Improving energy security: risk mitigation and management.

The important benefits listed above, may not be categorized as direct benefits to either the DES Utility or the building owner, however, they are the only two parties paying for DES development. Measuring and giving credit to the private sector for the positive contributions of DESs to the public sector, would help policy makers take advantage of synergistic eco-efficiencies.

5 Electricity Generation

5.1 Eco-efficiency Opportunity

In most of North America, the electrical grid faces the most strain during the hottest days of the year due to the electricity requirements for air conditioning. Cooling loads, however, can vary substantially throughout the day and because the highest electricity demand comes in the summer time during the hottest time (IESO 2010), the greatest benefit from conservation efforts will come from reducing energy demand at that time.

1 kWh saved during peak time is more significant than one saved at night. Exacerbating the problem, line losses are higher during peak times than low usage times. When the electricity transmission and distribution systems get hotter, the loss can be substantially higher than the average. In Ontario, the variation ranges from 5 % during low usage times to 25 % during peak hours according to an Ontario Hydro study. Taking into account the total losses from generation to delivery, saving 1 kW during peak times can reduce generating requirements by 1.47 kW (Ontario Hydro 2007).

> By reducing peak demand, the province can reduce its use of the less environmentally attractive resources that are called on when demand is high. In the long run, lower peak demand will mean less need for new generating facilities and transmission and distribution infrastructure, lowering costs for all Ontarians (Ontario Ministry of Energy).

Figure 3 shows the electricity demand of a 30 story glass office tower in Toronto for the year before the transition to district cooling and after. It is indicative of the savings that DES could provide. From the graph, it is clear when this particular building switched from running its own chillers to a DES; the

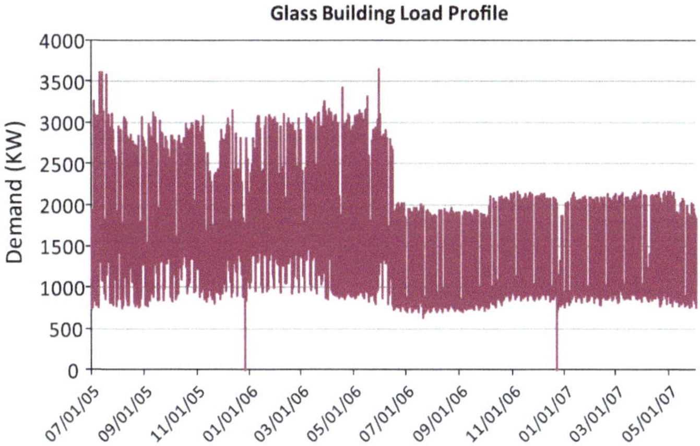

Fig. 3 Electrical load before and after conversion to DLWC for a glass building in Toronto (Toronto Building Manager 2011)

demand in electricity drops by roughly one third at the time of switchover. Reducing electrical demand translates into fewer gas fired generating stations needing to be built in populated areas.

5.2 Cost Savings

Electrically driven chillers for the most part generate more than 1 thermal unit of cooling energy for each unit of electricity. Rather than reference efficiencies in terms of percentages greater than 100, it is common practice to refer to chiller efficiencies in terms of Coefficient of Performance (COP). A fairly inefficient chiller with a COP of 3 produces 3 units of thermal energy for each unit of electrical energy. Assuming that chillers in the DES on average can produce a COP of 5 (requiring 0.1 kW_e of electricity per kW_{th}) compared to a BAU COP of 3 (requiring 0.3 kW_e of electricity per kW_{th}) (mostly due to scale efficiencies) at peak times, each kilowatt thermal (kW_{th}) of cooling demand translates to 0.2 kW_e of electricity demand reduction.

If DES utilities are able to capitalize on their size and generate and store chilled water at night (instead of producing it during the day), the peak time generation demand can reduce the cooling energy to effectively zero, saving 0.5 (kilowatt electrical) kW_e for each kW_{th} of cooling required.

In the case of free cooling by snow, lake water, or other natural sources, the peak time savings are also on the order of 0.5 kW_e for each kW_{th} of cooling since in both cases of chill storage and free cooling, only pumping energy for the water is required. Table 1 summarizes the possible load reductions.

For a contracted gas fired power plant in Ontario, each MW of electricity generation cost nearly \$1 M CDN (Pristine Power 2011). However, this cost is

Table 1 Peak reductions at generation source per kW_{th} cooling due to DES when compared to BAU

	District energy technology		
	Centrifugal chillers	Thermal storage	Free cooling
Peak reduction (at source)/kW_{th}	0.1–0.3 kW_e	~0.5 kW_e	~0.5 kW_e

strictly for technical costs. These costs could escalate quickly when projects are delayed or cancelled after contracts have been signed. Every MW of generation that can be avoided translates into significant savings.

5.3 Environmental Benefits

Reducing electricity use at peak times not only reduces the need to build more generating capacity, it potentially reduces atmospheric pollution and Greenhouse Gas (GHG) emissions.

Ontario, for example, uses a combination of nuclear and hydro power to satisfy the base load of electricity. As demand increases, the source of power tends to get dirtier. Figure 4 is a representative day of electricity generation in Ontario. As demand increases, high pollution and GHG emitting sources of generation come online. By using less energy during peak times (usually from noon to early evening), DES can reduce CO_{2e}, NO_x, and SO_x in the atmosphere.

Figure 4 shows a single day snap shot of the electrical output in Ontario. It is readily apparent that a base load is supplied by nuclear and hydro while natural gas

Fig. 4 Typical summer electrical output by source in Ontario (IESO)

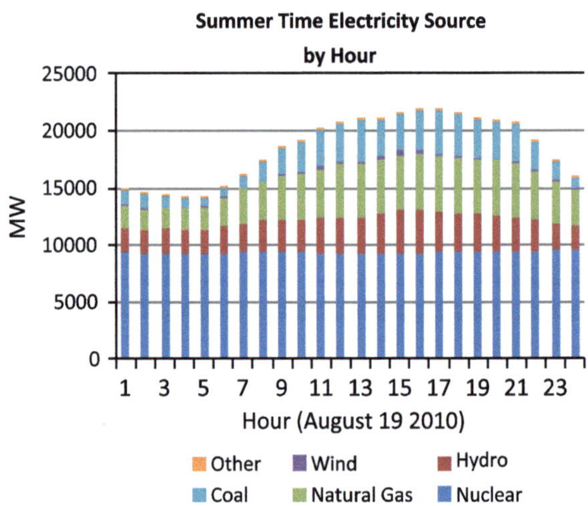

Table 2 NO_x and So_x output for various electricity generation forms (NRCAN GHGenius 2010)

	g/kWh				
	Coal	NG Boiler	NG Turbine	Nuclear	Hydro
NO_x	2.64	0.59	0.56	0.00	0.00
SO_x	4.96	0.01	0.01	0.00	0.00

is the first choice for marginal requirements. As the demand increases, coal starts to come online as a last resort given how dirty it is.

Table 2 shows the difference in NO_x and SO_x output between different forms of electrical generation. Ontario still requires coal to meet its demand at peak times. Coal fired power plants produce nearly 3 g of NO_x for each kWh of electricity and 5 g SO_x for each kWh of electricity. Conversely, nuclear and hydro do not add atmospheric pollution (during the operational phase of the life cycle). Using DESs to simply shift what time electricity is used for heating and cooling buildings could have a dramatic effect on pollutant emissions.

Although, studies have shown the correlation between NO_x, and SO_x with asthma and other respiratory illnesses (Lebowitz 1996; Detels et al. 1991), the cost savings from reducing emissions is difficult to calculate. Countries such as Canada, with government funded health care, can consider the cost savings associated with reducing the incidences of asthma and related respiratory illnesses as a bottom line benefit. A person with asthma in Ontario will on average cost the health care system twice what a person without the disease would cost (To 2007). If regulators and medical professionals can estimate the correlation between health care costs and the amount of pollution in the air, policy makers can begin to understand the dollar value of preventative versus reactionary health care costs.

In the same way that atmospheric pollution can be reduced, DESs can play a role helping countries meet GHG reduction targets. The authors conducted a study based on the electrical output in Ontario for 2009 and 2010; Table 3 shows that DES could decrease GHG emissions by as much as 145 g of CO_{2e} for each kWh of thermal cooling.

In many jurisdictions, there is no value assigned to GHG reduction. In these areas, DES utilities and building owners receive no economic benefit for the reduction in GHGs that a DES offers. Governments can encourage adoption of DES by providing an incentive equal in value to reduction initiatives. Alternatively, a fee or tax on emissions could incentivize connection to DESs.

Table 3 GHG reduction per KWh_{th} cooling (IESO, NRCAN 2010)

	District energy technology		
	Centrifugal chillers	Thermal storage	Free cooling
GHG reduction/kWh_{th}	67 g	35–145 g	145 g

Putting a price tag on GHG emissions and pollution reduction may not be popular in many jurisdictions but policy makers should recognize its merits. Deciding the value of each ton of GHGs reduced and applying that value to projects could spur innovation beyond DES.

6 Storm Water Management

6.1 Eco-efficiency Opportunity

DESs can free up rooftop space for green roofs. Traditionally, large buildings install thermal generating equipment such as boilers, chillers, and cooling towers in a penthouse mechanical room. These rooms diminish the ability to apply Low Impact Development techniques such as green roofs or rainwater harvesting for retaining storm water; water that runs into the city sewer system adding to the risk of Combined Sewer Overflows (CSO). By migrating to a DES connection building roof space is freed, enabling the application of low impact storm water containment strategies and government to capitalize on this synergistic opportunity.

6.2 Cost Savings

In the United States alone, \$44 billion (USD) is spent annually in CSO management (Montalto et al. 2007). Mitigating the risk of overflow can be a costly endeavor depending on local climate patters. Containment tanks and other end of the line solutions are effective but expensive solutions to CSO management. The local cost of CSO management needs to be considered, since it will vary from jurisdiction to jurisdiction, to evaluate the benefit in increasing rooftop green space that can be realized by connecting buildings to a DES. In order to determine the value of CSO mitigation, governments must weigh:

- The cost of standard CSO techniques
- The cost of land devaluation
- The public opinion on containment tanks and retention ponds.

Storm water containment is a difficult task given the variability of rain and flash flooding. As cities densify and the built urban form becomes less accommodating to storm water management, this will be more of a challenge for city planners. Standard CSO techniques may include ponds or containment tanks. These solutions are both costly and undesirable in neighbourhoods. When undesirable infrastructure is built, the land around it may depreciate in value which can adversely affect municipal tax revenue. Public opposition to CSO containment could also contribute to the cost of the infrastructure to municipalities.

Mismanagement of storm water can be costly. A single heavy storm in Toronto in August of 2005 cost the city $34 M (CDN) to repair damages. An additional $400 million (CDN) was paid out to private citizens from insurance companies due to the damage from this one storm (Riversides 2009a). The cost to mitigate the risk associated with CSO can quickly be paid back to municipalities by avoiding these types of events.

A cost metric of standard CSO containment can be established and compared with the mitigation provided per unit area of roof space by quantifying the cost and potential costs of a municipality's CSO containment strategy. An incentive could then be offered to building owners on a percentage of runoff reduction, since not all green roofs will have equivalent performance. If building owners are given an incentive that amounts to less than what a municipality would spend in containing an equivalent volume of storm water, storm water management costs to the municipality will be reduced.

6.3 Environmental Benefits

In addition to the cost of managing storm water, municipalities often contend with polluted water ways as a result of the runoff. The prevalence of cars in urban environments means that brake pad dust, tire wear fragments and motor oil are a normal part of storm water (Johnson 2009). New Jersey Department of Environmental Protection recognizes that "pollutant release is a serious threat to water quality" (NJDEP).

Excessive runoff transports toxins, picked up on roadways, into aquatic environments that can weaken or destroy plants and animals that depend on these (Riversides 2009b). In places, such as Toronto, the runoff overflows into the same body of water where potable water is drawn from. The effects of storm water present a serious concern, for public health and local ecology, that can be reduced by DESs.

7 Energy Security: Risk Mitigation and Management

7.1 Eco-efficiency Opportunity

DESs provide governments an opportunity to protect the reliability of local energy systems through conservation, diversity, flexibility, and availability. The International Energy Agency (IEA) defines energy security as "the uninterrupted physical availability at a price which is affordable, while respecting environmental concerns." With the growth of urban centers and the associated energy intensification required to provide essential services, governments are increasingly challenged to address issues of energy security. DESs can address factors, such as supply, price stability, and sustainability that contribute to greater energy security and independence.

The DES thermal grid which connects energy producers to end-users aids in conservation efforts. This connection allows waste heat from industrial or power generating processes to be used for residential and commercial heating. By utilizing waste heat sources, DESs reduce the amount of fuel burned for space heating and improve the efficient use of fossil fuels. Figure 5 illustrates that using waste heat from conventional power production, a concept known as CHP, can increase system efficiency and reduce fuel input from 147 to 100 units, a 30 % reduction. It is estimated that 61 % (OEE 2008) of building energy usage in Canada is used for space heating and cooling and water heating; a 30 % reduction fuel used in buildings is substantial. Waste and renewable fuels are available for heating and since fossil fuels are finite, there is value to reserving this precious resource for applications that have no other alternative such as pharmaceuticals and medical devices and equipment.

DESs can expand the diversity of fuel types used for heating and cooling by taking advantage of local fuel sources that would otherwise remain unused. Fuels such as biomass (wood chips, sawdust, straw), geothermal, biogas, and municipal waste are difficult to manage on a small scale—largely due to handling issues. The availability of local fuel sources and the ability to use them reduces the reliance on supplies from countries or jurisdictions that may be adversely affected by war, politics, or natural disaster. According to Sandor Boyson, research professor and co-director of the Supply Chain Management Center, "The longer the supply chain, the more that can go wrong and the more it costs with high gas prices."

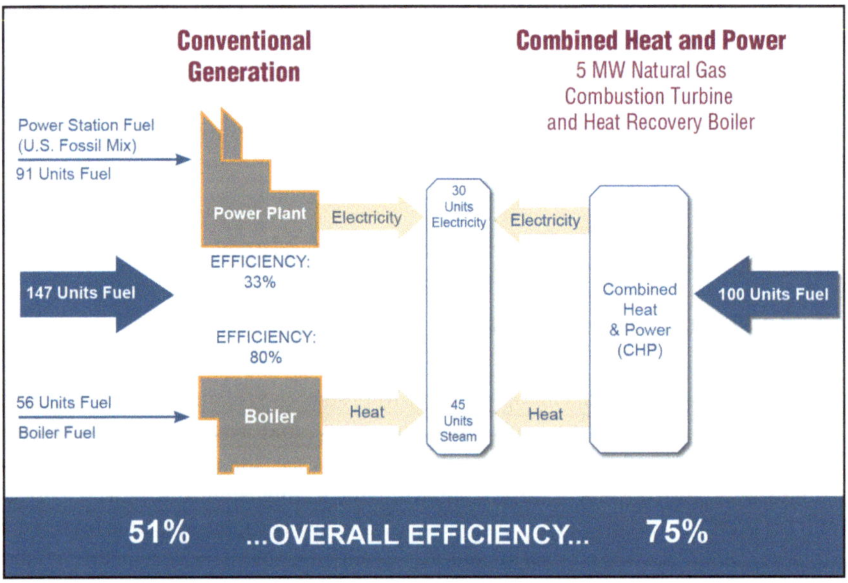

Fig. 5 Efficiency in combined heat and power systems (EPA 2012)

(University of Maryland, date unknown). In other words, a diverse fuel mix that incorporates local sources increases energy independence and reduces risk of supply interruptions and price instability.

The thermal grid and the centralized nature of energy production facilities allow DESs to support fuel switching and the implementation of state-of-the-art technologies. These qualities are referred to as fuel flexibility and technology flexibility; both improve the ability to optimize energy production in terms of cost and efficiency. The scale and centralized nature of DESs allow multiple fuel sources and/or technologies to be integrated at one location at a lower cost, than at many, in each individual building. In some cases certain fuels and technologies cannot even be applied on an individual building scale.

7.2 Cost Savings

The risk of not addressing energy security is real and can have a significant impact on the economy. Costs can be examined from at least two perspectives:

- Cost of interrupted to fuel/energy supply
- Cost of the inflexible nature energy production.

The recent power outage in India of July 2012, the North-eastern Blackout of August 2003, and the North American Ice Storm of January 1998 are only a few reminders of our reliance on energy. An estimate of the economic impact of the North-eastern Blackout of 2003 is in the range of $7.0 billion dollars (USD) from food spoilage, lost production, wages, etc. with the loss of 61,200 MW (ECLON 2004). Using the above figures, the economic impact of an energy supply interruption could be on the order of magnitude of $114,000 (USD)/MW of electrical supply loss (over the outage period). A catastrophic failure, similar in duration, of the W.A.C. Bennett Dam in B.C. at 2,730 MW or the Adam Beck I and II (Niagara River) in Ontario at 2,278 MW can have an economic impact in the order of $228 M (USD).

The value of mitigating energy (thermal and electrical) supply interruptions and pricing is difficult to estimate. As a gross estimate, municipalities could estimate the value of lost productivity as a function of the average GDP. This may help put into perspective the cost of each hour of unavailable electricity and thermal energy.

The inability of buildings to retrofit existing equipment to use alternate or renewable fuels or to implement more efficient technologies can reduce the competitiveness and robustness of an economy. Many industries, businesses (especially small ones), and households would encounter financial difficulties dealing with the consequences of a sharp increase in energy prices. Fluctuating energy prices can have a negative effect on many industries such as manufacturing, mining, transportation, forestry, and agriculture resulting in unemployment, loss of skilled labour and high paying jobs, as well as higher priced food and consumer goods resulting in a decreased standard of living and dampening of economic activity.

The value of being prepared for changes in the future, of being flexible, and of having a diversity of local fuel sources can be estimated by exploring the capital cost to modernize existing building heating and cooling systems for a group of buildings compared to a large scale DES. Additional factors to examine can include depreciated building value and the exposure to utility cost fluctuations of relying on a single fuel source for example, electric baseboard heating).

7.3 Environmental Benefits

Retrofitting hundreds of small boilers to use waste wood, bio-fuels, or solar energy would be much more expensive than modifying a single energy centre. The ability to switch fuels in a cost-effective manner means a higher likelihood of space heating and cooling needs being met by renewable sources.

Oujé-Bougoumou, an early adopter of DES in Canada has found that the presence of DES has displaced conventional energy sources such as fossil fuel and raised community awareness to environmental issues (Ouje 2012).

Energy is integral to modern living and the impact of energy supply interruptions is real and tangible. Through conservation, diversity, and flexibility, DESs contribute to greater energy security and independence.

8 Public Benefits of District Energy

8.1 NIMBY Infrastructure Reduction

Infrastructure, such as subways, roads, and clean water distribution, generally increase the value of land, however, certain forms of infrastructure, though necessary, decrease the value of land or add no value at all. NIMBY (Not-In-My-Back Yard) infrastructure is considered to be any infrastructure that is necessary yet undesirable. Examples of this are train tracks, hydro lines, and electricity generating stations. Although all are invaluable to society, opposition to building these forms of infrastructure is often fierce and the land around them rarely prized.

Building NIMBY infrastructure is costly and unpopular. It is difficult to overcome public opposition to where NIMBY infrastructure gets located. In Ontario, the difficulties in locating gas fired power plants were recently highlighted by the high-profile cancellation of a plant in the city of Mississauga. Despite having broken ground, the government was forced to cancel the project mid-election at a cost of $190 M (CDN) (Leslie 2012). Much of this could have been avoided if a DES was in place to mitigate the need for such infrastructure. Reducing the construction of NIMBY infrastructure has benefits both environmentally and politically.

8.2 Benefits Beyond Price

In evaluating eco-efficiency opportunities, in some cases, there are direct costs; for example the cost of NIMBY infrastructure reduction can be calculated. Not all benefits have obvious or cost savings. For example reducing the incidence rate of asthma and other pollutant related diseases could reduce the burden on state funded health care systems but it is difficult to put a price tag on the value of a healthy person or improved quality of life. Important questions to ask when developing sustainability plans include:

- What is it worth to the city/town to avoid living next to an electricity generator?
- What is it worth of reducing air emissions? If a price cannot be put on health, can a price be put on net emissions?
- How much has it cost to the economy, historically, when the electricity grid fails? How much financial risk is the municipality willing to accept?

There are challenges to limiting evaluation metrics to only measurable cost but it does not diminish the value of considering externalities such as clean air and water, improved health, community engagement, and public opinion.

By reaching outside of the direct cost benefits and including ancillary benefits to the analysis, the authors hope to create a broader more accurate evaluation of DES and the role it plays in reducing demands on energy infrastructure and consumption.

9 Policy Recommendations

Municipal, regional, and federal governments should take advantage of the real dollar savings that DES affords. This can only be done by identifying, understanding, and quantifying the costs and eco-efficiencies that DESs could save municipalities. Table 4 summarizes the opportunities and benefits that have been

Table 4 Eco-efficiency opportunities of district energy systems

Eco-opportunity	Benefit	Metric/Key performance indicators
Electricity generation	Demand reduction	kW demand reduction/kW generating capacity reduction Air pollution reduction
Storm water management	Runoff reduction	%/$ reduction in CSO needs or mitigation Water pollution reduction
Economic risk mitigation	Electrical grid stability	% reduction of blackouts
	Energy price stability	Reduce volatility
	Fuel flexibility	Minimize time to convert primary fuel sources

identified. The metrics and key performance indicators can be used as a guide for developing sound public policy surrounding DES. Without strong public policy to lead the way, the full environmental benefits of DES will be difficult to realize.

Beyond financial incentives or economics, the greatest hurdle to developing DESs, as in any business, is getting customers—without customer buildings, there can be no DES development. Even with a business case, signing customers is challenging because the status quo or engrained industry practices are difficult to overcome. People are generally adverse to change current business practices, especially when a proven method is achieving good results. Exacerbating the problem, many buildings are built and developed by a separate entity which owns and operates the building leaving little incentive to seek out efficiencies, reduce GHG emissions, or address energy security. All levels of government can do more to encourage DES development and bridge the gap toward affecting change.

9.1 Lead by Example

Public buildings should lead the way and be the first to connect to DES. In the absence of a DES in the municipality, government buildings should declare their intentions to connect at the earliest opportunity and even take steps to becoming DE Ready. DE Readiness means that buildings are able to convert to DES with minimal effort which means ensuring that buildings are designed with the following characteristics:

- Main mechanical room located in the basement or ground floor level,
- A centralized water-based (hydronic) heating and cooling system,
- Lowest hot water return temperatures and highest chilled water return temperatures as possible,
- High density or energy usage buildings, situated in close proximity to one another with a variety of usages.

There is often strong resistance to being the first to use a new or different technology or system. By leading the pack, government buildings can reduce the apprehension of other building owners by providing an example to inspire.

9.2 Create a Customer Base

One of biggest problems for DE Utilities is the absence of a ready customer base. Municipalities can take North Vancouver's example and mandate DE Readiness in all buildings before issuing permits. The next step would be to mandate DE connections in areas with an existing DES.

9.3 Minimum Performance Requirements

The authors also recommend that municipalities or regional governments establish minimum performance requirements in their jurisdictions. These performance metrics should be based on new state of the art technologies as opposed to existing building codes. Performance metrics should clearly include maximum allowable energy use, peak energy demand, and storm water runoff.

10 Further Research

To help municipalities and regional governments get the most out of eco-efficiency opportunities presented by DESs, policy makers must quantify their costs in managing storm water, securing electrical capacity, and addressing energy security in order to realize the potential synergies and cost savings.

Mitigating storm water run-off is a continual and potentially increasing problem for municipalities with the changing climate. As municipalities grow, the sewer system must deal with the corresponding increase in run off. The cost of dealing with high run off coefficients will vary from location to location and budgeted for by city engineers. Knowing the cost and benefits of various storm water runoff management strategies will help sustainability managers design appropriate incentives to reduce overall costs to the municipality.

Electrical utilities are common and the cost of each additional kW of new generation (for each technology) is fairly well studied, however, societal costs are of-ten ignored and focus only on design and construction costs. The cost of overcoming public outrage to this type of infrastructure should be taken into account when determining an overall cost per kW of additional generation. This will help policy makers appreciate the cost of cancelled plants, relocations, and other risks that are often difficult to budget.

11 Conclusion

It is critical to recognize the ancillary benefits of energy reduction techniques and the multiple benefits of DES to identify eco-efficiency opportunities. By quantifying the cost savings of avoided NIMBY infrastructure, recognizing the benefits of emissions reduction, and valuing energy security on the local level, municipalities and regional governments can make sustainable initiatives more cost effective for all parties—and in doing so reduce infrastructure costs and environmental impact.

Recognizing how systems interact with each other and the benefits they provide is an essential part of developing sustainable systems. The recognition of cost reduction and environmental benefits will allow governments at multiple levels to optimize their municipal service and pollution reduction strategies.

References

CDEA (2011) Is district energy a new technology, Canadian District Energy Association. https://www.cdea.ca/faq/district-energy-new-technology. Accessed 12 Jun 2012

Chan KT, Yu FW (2002) Part load efficiency of air-cooled multiple chiller plants. J Building Serv Eng Res Technol 23(1):31–41

Detels R et al (1991) The UCLA population studies of CORD: X. A cohort study of changes in respiratory function associated with chronic exposure to SOx, NOx, and hydrocarbons. Am J Public Health 81(3):350–359

ELCON (2004) Economic impacts of the August 2003 blackout. Electricity consumers resource council. Available via http://www.elcon.org/Documents/EconomicImpactsOfAugust2003 Blackout.pdf Accessed 05 Mar 2012

EPA (2012) Figure 5 courtesy of the United States Environmental Protection Agency. Used with permission

IESO (2010) Independent Electricity System Operator—hourly demand data was provided to authors for 2009 and 2010. Data is available upon request to IESO

IDEA (2012) International District Energy Association. Figures 1 and 2 courtesy of IDEA (Used with permission)

Johnson B (2009) Pollutant release: a 'dirty secret' in stormwater treatment. Stormwater research. Environ Sci Eng Mag (Summer 2009 edition)

Lebowitz MD (1996) Epidemiological studies of the respiratory effect of air pollution. In: Paoletti P, Costabel U (ed) Respiratory effects of air pollution. Eur Respir J 9:1029–1054

Leslie, Keith (2012) Taxpayers take $190 M hit for nixed gas plant. The Canadian Press, 17 July 2012

Montalto F et al (2007) Rapid assessment of the cost-effectiveness of low impact development for CSO control. Landscape Urban Plann 82:117–131

NRCAN (2012a) Avoid high speeds. Available via office of energy efficiency. http://oee.nrcan.gc.ca/cars-light-trucks/driving/fuel-efficient-driving-techniques/avoid-high-speeds/17834. Accessed 28 Aug 2012

NRCAN (2012b) Comprehensive energy use database Table 33. Available via office of energy efficiency. http://oee.nrcan.gc.ca/corporate/statistics/neud/dpa/tablestrends2/com_on_33_e_4.cfm?attr=0, Accessed 20 Apr 2012

NRCAN (2010) GHGenius v. 4. http://www.ghgenius.ca/ Accessed 28.08.2012

Ontario Hydro (2007) Distribution line loss study. EB-2007-0681, 15 Aug 2007

Ontario of Energy, "Smart Meters and Time of Use Prices", http://www.energy.gov.on.ca/en/smart-meters-and-tou-prices/

Ouje (2012) District heating system. available via Ouje Bougoumou, the place where people gather. http://www.ouje.ca/content/our-story/energy.php#benefits. Accessed 29Aug 2012

Pristine Power (2011) Pristine power announces award of a 20 year contract to York Energy Centre. Available via Newswire. http://www.newswire.ca/fr/story/298209/pristine-power-announces-award-of-a-20-year-contract-to-york-energy-centre Dec 2011

Riversides (2009a) Cost of mismanagement. Available via Toronto homeowner's guide to rainfall. http://www.riversides.org/rainguide/riversides_hgr.php?cat=1&page=78&subpage=79&subpage2=121. Accessed 27 Aug 2012

Riversides (2009b) The environmental cost of stormwater. Available via Toronto home-owner's guide to rainfall. http://www.riversides.org/rainguide/riversides_hgr.php?cat=1&page=78&subpage=90, Accessed 27 Aug 2012
Toronto Building Manager (2011) Source data for electrical usage provided to authors made anonymous by request. Data available upon request
University of Maryland (2012) Shorter supply chains and risk management pay off big. Available via Robert H. Smith School of Business Smith School News. http://www.rhsmith.umd.edu/news/stories/2011/SupplyChain.aspx, Accessed 31 Aug 2012

Environmental Issues in Vehicle Routing Problems

Ignacio Eguia, Jesús Racero, Jose Carlos Molina and Fernando Guerrero

Abstract In the last decade interest in environment preservation is increasing and environmental aspects play an important role in strategic and operational policies. Therefore, environmental targets are to be added to economic targets, to find the right balance between these two dimensions. Green logistics extend the traditional definition of logistics by explicitly considering external factors associated mainly with climate change, air pollution, noise, vibration and accidents. Among the logistical activities, the vehicle routing problem (VRP) is one of the most widely researched and has mainly focused on economic objectives, not considering explicitly environmental issues. In this chapter, a realistic variant of the VRP with heterogeneous vehicle fleets in which vehicles are characterized by different capacities and costs, has been considered and external costs have been estimated using international research projects, and have been included as part of a mixed-integer linear programming model to solve a realistic variant of the VRP. To solve medium to large-size VRP instances, heuristic approaches are necessary. An impressive number of heuristic have been proposed for the VRP in the literature. In this chapter, one heuristic is developed to find good solutions to the proposed eco-efficiency model: a savings heuristic when time windows are not considered. Since there are no instances for this problem variant, the algorithm is validated with benchmarking problems adapted from the literature, offering good solutions and quickness. The selection of eco-efficiency routes can help to reduce the emissions of air pollutants, noises and greenhouse gases, without losing competitiveness in transport companies.

I. Eguia · J. Racero (✉) · J. C. Molina · F. Guerrero
Escuela Superior de Ingeniería, Universidad de Sevilla, Camino de los Descubrimientos, Sevilla, Spain
e-mail: jrm@us.es

M. G. Erechtchoukova et al. (eds.), *Sustainability Appraisal: Quantitative Methods and Mathematical Techniques for Environmental Performance Evaluation*, EcoProduction, DOI: 10.1007/978-3-642-32081-1_10,

Keywords Vehicle routing problem · Sustainability · Externalities · Green logistics · Environmental costs

1 Introduction

Environmental issues can impact on numerous logistical decisions throughout the supply chain such as location, sourcing of raw material, modal selection, and transport planning, among others. Green logistics extends the traditional definition of logistics ("integrated management of all the activities required moving products through the supply chain at minimum cost") by explicitly considering other external factors associated mainly with climate change, air pollution, noise, vibration and accidents.

The logistical activities comprise freight transport, storage, inventory management, materials handling and all the related information processing. In this chapter, we study the design of routes in road freight transportation activities, which are significant sources of air pollution, noise and greenhouse gas emissions, with the former known to have harmful effects on human health and the latter, responsible for global warming. Thus, we analyze the well-known Vehicle Routing Problem (VRP) to estimate the effects of those environmental issues in this type of problems.

The VRP deals with founding the optimal routes of delivery or collection from one or several depots to a number of customers, while satisfying some constraints. The VRP plays a vital role in distribution and logistics. Huge research efforts have been devoted to studying the VRP since 1959 and thousands of papers have been written on several VRP variants. We refer to the survey by (Cordeau et al. 2007) for coverage of the state-of-the-art on models and solution algorithms.

The classical VRP tries to minimize the total distance travelled by a set of vehicles while satisfying the demand of a given set of customers and considering the assumption that each vehicle serves a single route during any planning period. This problem is a generalization of the well-known and widely studied Traveling Salesman Problem (TSP). The TSP is in mathematical terms a NP-hard combinatorial problem and therefore also the VRP is NP-hard. Different variations of the classical VRP have been proposed with the aim of approaching to real contexts, where some variables and constraints have been included.

When demand of all customers exceeds the vehicle capacity, two or more vehicles are needed. This implies that in the Capacitated Vehicle Routing Problem multiple Hamiltonian cycles have to be found such that each Hamiltonian cycle is not exceeding the vehicle capacity. Then the CVRP consists of designing a set of least-cost vehicle routes in such a way that: (1) each customer has a positive demand D_i which has to be fully satisfied once by exactly one vehicle; (2) all vehicle routes start and end at the depot, which has no demand; and (3) the vehicle fleet is homogeneous, i.e. each vehicle has an equal capacity of Q.

The Multi-Depot Vehicle Routing Problem (MD-VRP) is a generalization of the CVRP in which vehicles start from multiple depots and return to their original depots at the end of their assigned tours. If customers are dispersed around the depots, then the distribution problem can be modeled as several independent CVRP. However, if they are intermingled then a MD-VRP should be solved. Each depot is responsible for a number of customers who are served by the vehicles assigned to the depot.

The Vehicle Routing Problem with Time Windows (VRPTW) is the variant that has received the most attention in the literature by the practical importance of time windows. Time windows occur when customers require pick-up or delivery within a pre-specified service times, characterized by an early time and a late time. In addition to time windows for customers, it may be included also a limit on the total driving time, that is, the maximum time allowed for any vehicle due to contract regulations for drivers. In the literature a distinction is made between soft time windows that can be violated against a penalty-cost and hard time windows that cannot be violated. The VRPTW has been the subject of intensive research efforts for both heuristic and exact optimization approaches. An overview of the early published papers is given by (Solomon 1987).

The Vehicle Routing Problem with Pick-up and Delivering (VRPPD) considers that besides the deliveries to a set of customers, a second set of customers requires a pick-up, where deliveries and pickups can be made at any order. As a particularity of this problem, in the Vehicle Routing Problem with Backhauls (VRPB) it is assumed that each vehicle will first visit customers that require delivery (line-haul customers) before it visits the customers (suppliers) that require pick-up (backhaul customers). This arises from the fact that the vehicles are rear-loaded, and rearrangement of the loads on the tracks at the delivery points is not deemed economical or feasible (Toth and Vigo 2002).

The VRP with split deliveries (SD-VRP) is a variant where it is allowed that the same customer can be served by different vehicles if this will help to reduce the total route costs. It occurs when the sizes of the customer orders are as big as vehicle capacities.

In classical VRP, the planning period is a single day. In the case of the Period Vehicle Routing Problem (P-VRP), the classical VRP is generalized by extending the planning period to M days. For this variant, the following constraints must be considered: (a) each vehicle must have a defined capacity, (b) each customer has a known daily demand that must be completely satisfied in only one visit by exactly one vehicle, and (c) it is not necessary that the vehicle returns to the depot the same day it came out, but must return in a time period already defined.

In Stochastic VRP (S-VRP) is assumed that one or several components of the problem are random. There are three different kinds of S-VRP: (a) customers with a probability of presence or absence, (b) customers whose demand is a random variable, or (c) customers where the service time and travel time are random variables. In S-VRP, two stages are made for getting a solution. A first solution is determined before knowing the realizations of the random variables. In a second stage, a recourse or corrective action can be taken when the values of the random variables are known. When some data are random, it is no longer possible to require that all constraints be satisfied for all random variables, so the decision maker may either require the satisfaction of some constraints with a given probability or the incorporation into the model of corrective actions when a constraint is violated.

A variant of the VRP arises when a fleet of vehicles (limited or unlimited), characterized by different capacities, fixed costs and variable costs, is available for distribution activities. The problem is known as Heterogeneous Fleet VRP (HF-VRP). The HF-VRP with unlimited fleet, known as the Fleet Size and Mix VRP (FSM), was first proposed by (Golden et al. 1984) and it consists of determining, at the same time, the best fleet composition and the optimal routing of a fleet with an unlimited number of heterogeneous vehicles in order to serve a given set of customers with deterministic delivery demands, minimizing the total travel costs. The HF-VRP variant with a limited number of vehicles, called Heterogeneous VRP (HVRP), was proposed by (Taillard 1999) and it consists in optimizing the routes with the available fixed fleet. In both cases, fixed costs (F) and/or dependent routing costs (D) could be considered. As a result the following HF-VRP variants have been studied in the literature: (1) HVRPFD, (2) HVRPD, (3) FSMFD, (4) FSMF, and (5) FSMD. In Sect. 3, we present a survey on HF-VRP.

The classical objective function in VRP is minimizing the total distance travelled by all the vehicles of the fleet or minimizing the overall travel cost, usually a linear function of distance. This objective is widely used by researchers with homogeneous fleet and when vehicles are not allowed to remain at the depot.

The main objective of companies with a heterogeneous fleet and with less demand than capacity consists of determining the fleet composition minimizing the fixed and the variant costs. Some authors (Sniezek and Bodin 2002) argue that only considering total travel time or total travel distance in the objective function is not enough in evaluating VRP solutions, especially for heterogeneous fleets. Instead, they determine a Measure of Goodness, which is a weighted linear combination of many factors such as capital cost of a vehicle, salary cost of the driver, overtime cost and mileage cost. These costs are considered as internal or economic costs for transportation companies.

In this chapter, the VRP with realistic assumptions and a new objective function that accounts environmental issues is considered. Thus, an eco-efficiency model of the VRP with Heterogeneous Fleet and Time Windows (HF-VRPTW) is presented with a broader objective function that accounts not just for the internal costs (driver, fuel, maintenance,...), but also for external costs (greenhouse emissions, air pollution, noise,...). This model is solved using a heuristic algorithm. With this approach, transportation companies can have positive environmental effects by making some operational changes in their logistics system, selecting the most appropriate vehicles, determining the routes and schedules to satisfy the demands of the customers, reducing externalities and achieving a more sustainable balance between economic, environmental and social objectives.

The remainder of this chapter is organized as follows. In Sect. 2 the environmental impacts of transportation activities and the cost estimation of these externalities are analyzed. Section 3 reviews the literature on the HF-VRPTW and on incorporating environmental issues in VRP. Section 4 presents the proposed approach, a mixed-integer linear programming model and a heuristic algorithm to solve it. A numerical example will be explained in Sect. 5 and a real case application is presented in Sect. 6. Section 7 presents the results and discussion. The conclusions and recommendations for further work are given in Sect. 8.

2 Externalities in Transport

In the last decade interest in environment preservation is increasing and environmental aspects play an important role in strategic and operational policies. Therefore, environmental targets are to be added to economic targets, to find the right balance between these two dimensions (Dyckhoff et al. 2004).

Transport activities give rise to environmental impacts, accidents and congestion. In contrast to the benefits, the costs of these effects of transport are generally not borne by the transport users. Without policy intervention, these so called external costs are not taken into account by the transport users when they make a transport decision. Transport users are thus faced with incorrect incentives, leading to welfare losses.

The internalisation of external costs means making such effects part of the decision making process of transport users. According to the welfare theory approach, internalisation of external costs by market-based instruments may lead to a more efficient use of infrastructure, reduce the negative side effects of transport activity and improve the fairness between transport users.

Internalization of external cost of transport has been an important issue for transport research and policy development for many years in Europe and worldwide. Some authors (Bickel et al. 2006) focus their research on evaluating the external effects of transport to internalize them through taxation. As a result, decisions such as the selection of vehicle types, the scheduling of deliveries, consolidation of freight flows and selection of type of fuel, considering internal and external costs can help to reduce the environmental impact without losing competitiveness in transport companies.

In this chapter, we focus our attention on external costs associated with: greenhouse emissions, atmospheric pollutant emissions, noise emissions and accidents. These four components reflect 88 % of the total average external cost freight in the European Union, excluding congestion costs (INFRAS/IWW 2004). The evaluation of each component of the external costs applied to the Spanish transport setting is based on the European study (INFRAS et al. 2008).

Climate change or global warming impacts of transport are mainly caused by emissions of the greenhouse gases: carbon dioxide (CO_2), nitrous oxide (N_2O) and methane (CH_4). The main cost drivers for marginal climate cost of transport are the fuel consumption and carbon content of the fuel. For internalization purposes the estimated external costs of CO_2 emissions can be factored into the total well-to-wheel greenhouse gas emissions per litre of fuel used by multiplying the grams of CO_2 per litre with the external costs per gram of CO_2 emitted.

The recommended value for the external costs of climate change for year 2010, expressed as a central estimate is 25€/ton.CO_2. The total well-to-wheel CO_2 emissions per unit of fuel, also called emission factor, is estimated in 2.67 kg of CO_2 per litre of diesel.

Air pollution costs are caused by the emission of air pollutants such as particulate matter (PM), NO_x and non-methane volatile organic compounds (NMVOC). Emissions of a road vehicle depend on vehicle speed, fuel type and the related combustion technology, the load factor, vehicle size, the driving pattern and the geographical location of the road. For internalization purposes the estimated external costs of each pollutant emissions can be obtained by multiplying the grams of the pollutant per kilometer travelled with the external costs per gram of pollutant emitted.

The recommended air pollution costs for each pollutant in Spain (emissions 2010, in €2000/ton of pollutant) are: $NO_x = 2{,}600$; NMVOC = 400; $PM_{2.5} = 41{,}200$; $PM_{10} = 16{,}500$, using PM in outside built-up areas. The ratio €2010/€2000 is fixed to 1.323. The estimation of pollutant emissions from road transport are based on the Tier 2 methodology (EMEP/EEA 2010). This approach considers the fuel used for different vehicle categories and technologies according to emission-control legislation.

Noise costs consist of costs for annoyance and health. For the estimation of noise costs data on the number of exposed people is needed. In road transport the sound emitted is mainly made up by the sound of the propulsion system and the sound of rolling. The ratio of both sources depends on the speed of the vehicle, the vehicle type, the kind of tyres, the vehicle's state of maintenance, the slope of the road and the kind of surface.

The recommended noise costs based on (INFRAS/IWW 2004) for Heavy-Duty Vehicles are in a range from 0.25 to 32 (in €2000/ton-km) considering different vehicle categories, countries and traffic situations, with a mean value of 4.9.

External accident costs are those social costs of traffic accidents which are not covered by risk oriented insurance premiums. The most important costs in road transport are besides vehicle kilometers, vehicle speed, type of road, drivers' characteristics, traffic speed and volume, time of day (day/night) and interaction with weather conditions.

The recommended accident costs also based on (INFRAS/IWW 2004) for Heavy-Duty Vehicles are in a range from 0.7 to 11.8 (in €2000/ton-km), with a mean value of 4.75.

In this chapter, the designed routes will employ all of these average costs and emission factors, multiplying these parameters by the respective distance travelled, load carried or fuel consumed in each route.

3 Literature Review

3.1 VRP with Heterogeneous Fleet

As mentioned in Sect. 1, several variants on VRP inspired by real world applications were proposed over the years. Our interest in this chapter relies on the heterogeneous fleet VRP (HF-VRP), instead of the homogeneous fleets. In industry, fleets are rarely homogeneous and companies are incorporated vehicles with different features over the time (Hoff et al. 2010). As a result, standing and running costs depend on each vehicle according to depreciation level or usage time of the fleet.

The first HF-VRP variants were the FSM problems, proposed by (Golden et al. 1984). These authors suggested two heuristics where the first one is based on the savings algorithm of (Clarke and Wright 1964), and the second one use a giant tour schema. They proposed 20 instances with 12–100 customers and 3–6 vehicle types with an unlimited number of each one. Instances 1–12 have 12 or 30 customers. Instances 13–20 are larger, with 50–100 customers. This second set of instances is considered as benchmark in FSM problems.

The another HF-VRP variants were the HVRP problems, first introduced by (Taillard 1999) who presented a heuristic column generation method. This method starts by solving the homogeneous vehicle routing problem for each vehicle type using a tabu search algorithm. The routes obtained are stored in a set of possible final routes and then, through a process of successive iterations, these routes are extracted and combined into a partial solution to the criterion of non-repetition of customers. The final set of routes is obtained by solving a SP problem minimizing the total costs and ensuring that each customer is served by only one route. This method is tested for the FSMF and FSMD on the second set of instances of (Golden et al. 1984), and also for the HVRPD adding a limited number for each type of vehicle. This set of 8 instances is considered as benchmark in HVRP problems.

As the HF-VRP is a special case of the classical VRP, these problems are NP-Hard and therefore heuristic algorithms are suitable approaches to obtain high-quality solutions in an acceptable computing time. To the best of our knowledge, no exact algorithms have been developed for any variant of HF-VRP.

Some mathematical formulations have been presented in the literature. Gheysens et al. (1984) formulated the FSMF using a three-index binary variables x_{ij}^k to represent if a vehicle of type k travels directly from customer i to customer j. Golden et al. (1984) proposed a similar formulation for the FSMF but using the Miller-Tucker-Zemlin constraints for the TSP to avoid sub-tours (Miller et al. 1960). Yaman (2006) also described six different formulations for the FSMF using these binary variables. Another type of formulations for FSMF is based on the Set Partitioning (SP) model of the VRP and associates a binary variable with each feasible route (Balinski and Quandt 1964). Mathematical formulations for the FSM with Time Windows were described in (Dell'Amico et al. 2006) and (Bräysy et al. 2008).

Although no exact algorithms have been proposed for the FSM, some lower bounds were presented. Golden et al. (1984) proposed some lower bounds for the FSMF problem when there are symmetric distances and triangle inequality between customers. Yaman (2006) also proposed lower bounds for the FSMF based on cutting-plane techniques applied to the six mathematical formulations. Choi and Tcha (2007) extended the lower bounds to all the variants of FSM problems (FSMFD, FSMD and FSMF) based on a SP formulation and using a column generation technique. Pessoa et al. (2009) developed a Branch-Cut-and-Price algorithm to obtain lower and upper bounds for the three FSM variants. Baldacci and Mingozzi (2009) proposed a SP algorithm with bounding procedures based on linear and Lagrangian relaxation to solve the five variants of HF-VRP mentioned in Sect. 1.

All solutions approaches presented in the literature are heuristic algorithms, and they usually are adaptations or extensions of heuristics applied to classical VRP variants. In this way, since the late 90s, metaheuristic approaches are applied to find high-quality solutions of HF-VRP as well.

After (Golden et al. 1984) developed two constructive heuristics to adapt the savings and giant-tour for solving the FSMF problem, some other authors also proposed constructive heuristics (Gheysens et al. 1984; Salhi and Rand 1993). Renaud and Boctor (2002) proposed an extension of the sweep algorithm for classical VRP, to solve the FSMF problem.

Considering metaheuristic approaches, some authors implemented heuristic procedures based on Genetic Algorithms (GA). Ochi et al. (1998) developed a hybrid algorithm that combines GA and Scatter Search to solve the FSMF. Also, (Liu et al. 2009) proposed a GA with a local search to solve the FSMF and FSMD. Lima (2004) solved the FSMF using a Memetic Algorithm (MA). Prins (2009) developed two heuristic procedures based on a MA for all the variants of the FSM problem and for the HVRPD.

Tabu Search (TS) is one of the most extended metaheuristics applied to VRP. Some TS approaches were proposed for solving FSMF and FSMD. Gendreau et al. (1999) developed a TS algorithm based on the GENIUS neighborhoods and an adaptive memory procedure. Lee et al. (2008) proposed a TS algorithm combined with a SP approach. Brandao (2009) proposed a deterministic TS with different procedures to generate the initial solutions.

Recently, new metaheuristics have been applied to FSM. Imran et al. (2009) developed a Variable Neighborhood Search (VNS) algorithm for all FSM variants. Penna et al. (2011) presented a hybrid heuristic based on the Iterated Local Search metaheuristic which uses a VNS procedure in the local search phase.

Prins (2002) developed a heuristic for solving the HVRPD by extending a series of VRP classical heuristics and incorporating a local search procedure based on the Steepest Descent Local Search and TS.

Tarantilis et al. (2003) and later Tarantilis et al. (2004) developed a list-based threshold accepting algorithm (denoted by LBTA) and a backtracking adaptive threshold accepting algorithm (denoted by BATA) for solving the HVRPD. The idea of this class of algorithms is to allow moves toward solutions with higher objective function values (uphill moves) in order to escape from local minimums.

Both methods start with an initial solution generated by a constructive heuristic. The algorithms seek feasible solutions in the neighborhood (local search) and are compared to a list storing the T best threshold values. In LBTA, the list of T values is used during the search, while in BATA the list of T values is allowed to increase during the search.

Two years later, (Li et al. 2006) developed a similar algorithm called HRTR to solve HVRPD and HVRPFD problems, based on the algorithm Record-To-Record (RTR), a deterministic variant of the Simulated Annealing metaheuristic. The algorithm accepts neighbor solution with less objective function value than actual ones plus a deviation, avoiding a local minimum. Then, using a local search (downhill moves), the algorithm looks for new global minimums.

Finally, some heuristics were developed for the heterogeneous fleet VRP with Time Windows. Liu and Shen (1999) proposed a two-phase algorithm to solve the FSMFTW problem. In the first phase, a savings algorithm evaluates the insertion of complete routes in all the possible insertion places of the other routes, taking into account the time windows. In the second phase, intra-route and inter-route exchanges are performed to improve the best solutions found during the first phase. Computational results were performed on a set of 168 test instances derived from the (Solomon 1987) VRPTW test set. This set of 168 instances is considered as benchmark in FSM-TW problems.

Dullaert et al. (2002) extended to FSMFTW the sequential insertion algorithm proposed by (Solomon 1987) incorporating the (Golden et al. 1984) modified saving expressions. Dell'Amico et al. (2006) proposed a ruin-and-recreate metaheuristic approach using a parallel insertion procedure. Bräysy et al. (2008) proposed a deterministic annealing metaheuristic in three phases. These three heuristics has been tested on the 168 instances for solving the FSMFTW problems.

A HF-VRP survey with all the five variants mentioned in this chapter can be found in (Baldacci et al. 2008). Also a survey on industrial aspects of combined fleet composition and routing in maritime and road-based transportation heterogeneous fleet was performed by (Hoff et al. 2010).

As a result of this literature review, we consider that the HVRP is less studied than the FSM, and there are no benchmarking tests on HVRP-TW.

3.2 VRP and Sustainability

VRP studies have been benefiting green logistics from its origin, as early as their introduction by (Dantzig and Ramser 1959). This contribution has always been completely implicit and researchers have often been totally unaware of the beneficial connotations of their works on the environment.

While VRP aims at minimizing distances and total assigned vehicles, it is satisfying green transportation requirements by reducing the amount of fuel and consequently reducing the CO_2 emissions from road transportation.

The influence of VRP is not limited to minimizing travel distance and vehicle numbers, there are green transportation factors that could be considered in a VRP model and have been studied during the past few years.

The contribution of VRP to green logistics has its origins in studies of (Sbihi and Eglese 2007) and PhD dissertation of (Palmer 2007).

In a working paper for Lancaster University Management School, Sbihi and Eglese (2007) reviewed the literature related to vehicle routing in order to find the relationship between vehicle routing and scheduling problems (VRSP) and green logistics. They couldn't find much literature that links VRP models with the Green Logistics issues, but they argued that reduction in total distance would provide environmental benefits due to the reduction in fuel consumption and the consequent air pollutants.

Palmer (2007) suggested an integration of elements from transportation planning and environmental modeling combined with logistics based vehicle routing techniques for freight vehicles and investigated the role of speed in reducing CO_2 emissions under various scenarios and time windows settings. They developed a computer based vehicle routing model that calculates the overall amount of CO_2 emitted from road journeys, as well as time and distance.

Following in this emerging area, a number of studies taking account environmental considerations in their objective functions were published. The first important paper is the "Pollution Routing Problem (PRP)" by (Bektas and Laporte 2011). They defined the PRP as a variant of the VRP, with or without time windows, using a comprehensive objective function which measures and minimizes the cost of GHG emissions along with operational costs of drivers and fuel consumption. They also analyzed and compared between various performance measures of vehicle routing, such as distance, load, emissions and costs evaluated through a variety of objective functions.

Xiao et al. (2012) contemplated the Fuel Consumption Rate (FCR) as a load dependant function, and added it to the classical VRP to extend it with the objective of minimizing fuel consumption. They presented a mathematical optimization model to formally characterize the FCR considered CVRP (FCVRP) as well as a string based version for calculation. The results of the experiments showed that the FCVRP model can reduce fuel consumption by 5 % on average compared to the CVRP model.

Erdogan and Miller-Hooks (2012) introduced the Green Vehicle Routing Problem (G-VRP) as a mixed-integer linear program. They developed techniques to aid organizations with alternative fuel-powered vehicle fleets in overcoming difficulties that exist as a result of limited vehicle driving range in conjunction with limited refueling infrastructure. These techniques seek a set of vehicle tours that minimize total distance traveled to serve a set of customers while incorporating stops at Alternative Fueling Stations (AFSs) in route plans so as to eliminate the risk of running out of fuel. Given a complete graph consisting of vertices representing customer locations, AFSs, and a depot, the G-VRP seeks a set of vehicle tours with minimum distance each of which starts at the depot, visits a set of customers within a pre-specified time limit, and returns to the depot without exceeding the vehicle's driving range that depends on fuel tank capacity. Each tour may include a stop at one or more AFSs to allow the vehicle to refuel en route.

Ubeda et al. (2011) presented a case study to show how the introduction of green practices into the daily decision-making process in a transportation company can simultaneously meet efficiency and environmental objectives. They incorporated the CO_2 emissions in the objective function of the CVRP with backhauls and maximum allowable driving time.

Figliozzi (2010) introduced a different problem: the minimization of emissions and fuel consumption as the primary or secondary objective, creating the Emissions VRP (EVRP). He considered time windows and capacity constraints as well as time-dependent travel time. The chapter deals with a static problem, and the dispatcher is assumed to know the impact of congestion on travel speeds. The amount of emissions is a function of travel speed and it will depend of the departure time in each node.

Maden et al. (2010) considered a VRSP with time Windows in which Speedy depends on the time of the travel. They described a heuristic algorithm to solve the problem. Jabali et al. (2009) considered a similar problem but estimated the amount of emissions based on a nonlinear function of speed and other factors, finding the optimal speed with respect to emissions. Kara et al. (2007) introduced the Energy-Minimizing VRP, an extension of the VRP where a weighted load function is minimized, trying to minimize the energy consumed.

The above-discussed studies have been published recently. This shows that the topic is at its beginning and is still too attractive and demanding.

4 A Sustainable VPR Approach

4.1 An Eco-efficiency Model for HVRPTWB

The problem presented in this chapter is an extension of the classical Capacitated Vehicle Routing Problem, including Time Windows and Backhauls, and a Heterogeneous Fleet with different vehicles and fuel types (HVRPTWB). The following assumptions are stated about the problem: (a) known fleet size, (b) heterogeneous fleet, with different vehicle capacities, fuel consumptions and categories, (c) single depot, (d) deterministic demand, (e) oriented network, (f) time windows, (g) maximum driving time, and (h) backhaul nodes. The main contributions of this chapter deal with formulating a mathematical model of the HVRPTWB and with considering external and social impacts as part of the internal costs of the company. Then the overall objective is to minimize the total cost that is composed of internal costs (cost of drivers, energy costs, fixed cost of vehicles–depreciation, inspection, insurance-, maintenance costs and toll costs) and external costs (climate change, air pollution, noise and accidents).

The HVRPTWB is defined on a graph $G = \{N, A\}$ with $N = \{0,1,\ldots,t, t+1,\ldots,n\}$ as a set of nodes, where node 0 represents the depot, nodes numbered

1 to t represent delivery points and nodes numbered $t + 1$ to n represent supply points (backhauls), and A is a set of arcs defined between each pair of nodes. A set of m heterogeneous vehicles denote by $Z = \{1, 2,\ldots,m\}$ is available to deliver the desired demand of all customers from the depot node and then to pick-up the inbound products from the supply and return to the depot node. The constructing routes of each vehicle must meet the following constraints: no vehicle carries load more than its capacity, each customer and supplier is visited within its respective time window, customers are not visited after any suppliers and no vehicle exceeds the maximum allowable driving time per day.

We adopt the following notation:

D_i load demanded by node $i \in \{1,\ldots,t\}$ and load supplied by node $i \in \{t + 1,\ldots,n\}$
q^k capacity of vehicle $k \in \{1,\ldots,m\}$
$[e_i, l_i]$ earliest and latest time to begin the service at node i
s_i^k service time in node i by vehicle k
d_{ij} distance from node i to node j $(i \neq j)$
t_{ij} driving time between the nodes i and j
T^k maximum allowable driving time for vehicle k.

Our formulation of the problem uses de following decision variables:

x_{ij}^k binary variable, equal to 1 if the vehicle $k \in \{1,\ldots,m\}$ travels from nodes i to j $(i \neq j)$
y_i^k starting service time at node $i \in \{0,1,\ldots,n\}$; y_0^k is the ending time
f_{ij}^k load carried by the vehicle $k \in \{1,\ldots,m\}$ from nodes i to j $(i \neq j)$.

Constraints of the model are as follows:

$$\sum_{j=1}^{n} x_{0j}^k \leq 1 \quad (k = 1,\ldots,m) \tag{1}$$

$$\sum_{\substack{j=0\\ j\neq i}}^{n} x_{ij}^k - \sum_{\substack{j=0\\ j\neq i}}^{n} x_{ji}^k = 0 \quad (k = 1,\ldots,m; \quad i = 1,\ldots,n) \tag{2}$$

$$\sum_{k=1}^{m} \sum_{\substack{j=0\\ j\neq i}}^{n} x_{ij}^k = 1 \quad (i = 1,\ldots,n) \tag{3}$$

$$\sum_{i=1}^{t} D_i \sum_{\substack{j=0\\ j\neq i}}^{n} x_{ij}^k \leq q^k \quad (k = 1,\ldots,m) \tag{4}$$

$$\sum_{i=t+1}^{n} D_i \sum_{\substack{j=0\\ j\neq i}}^{n} x_{ij}^k \leq q^k \quad (k = 1,\ldots,m) \tag{5}$$

$$\sum_{k=1}^{m}\sum_{i=t+1}^{n}\sum_{j=1}^{t}x_{ij}^{k}=0 \tag{6}$$

$$\sum_{k=1}^{m}\sum_{j=t+1}^{n}x_{0j}^{k}=0 \tag{7}$$

$$y_i^k+s_i^k+t_{ij}\le y_j^k+T^k(1-x_{ij}^k)\quad (i=1,\ldots,n;\quad j=0,\ldots,n;\quad j\ne i;\quad k=1,\ldots,m) \tag{8}$$

$$t_{0j}\le y_j^k+T^k(1-x_{0j}^k)\quad (j=1,\ldots,n;\quad k=1,\ldots,m) \tag{9}$$

$$e_i\le y_i^k\le l_i\quad (i=1,\ldots,n;\quad k=1,\ldots,m) \tag{10}$$

$$y_0^k\le T^k\quad (k=1,\ldots,m) \tag{11}$$

$$\sum_{k=1}^{m}\sum_{\substack{j=0\\ j\ne i}}^{n}f_{ji}^k-\sum_{k=1}^{m}\sum_{\substack{j=0\\ j\ne i}}^{n}f_{ij}^k=D_i\quad (i=1,\ldots,t) \tag{12}$$

$$\sum_{k=1}^{m}\sum_{\substack{j=0\\ j\ne i}}^{n}f_{ij}^k-\sum_{k=1}^{m}\sum_{\substack{j=0\\ j\ne i}}^{n}f_{ji}^k=D_i\quad (i=t+1,\ldots,n) \tag{13}$$

$$f_{ij}^k\le(q^k-D_i)x_{ij}^k\quad (i=0,\ldots,t;\quad j=0,\ldots,n;\quad j\ne i;\quad k=1,\ldots,m) \tag{14}$$

$$D_jx_{ij}^k\le f_{ij}^k\quad (j=1,\ldots,t;\quad i=0,\ldots,n;\quad i\ne j;\quad k=1,\ldots,m) \tag{15}$$

$$D_ix_{ij}^k\le f_{ij}^k\quad (i=t+1,\ldots,n;\quad j=0,\ldots,n;\quad j\ne i;\quad k=1,\ldots,m) \tag{16}$$

$$f_{ij}^k\le(q^k-D_j)x_{ij}^k\quad (j=t+1,\ldots,n;\quad i=0,\ldots,n;\quad i\ne j;\quad k=1,\ldots,m) \tag{17}$$

Constraints (1) mean that no more than m vehicles (fleet size) depart from the depot. Constraints (2) are the flow conservation on each node. Constraints (3) guarantee that each customer and supplier is visited exactly once. Constraints (4) and (5) ensure that no vehicle can be overloaded. Constraint (6) guarantees that customers are not visited after any suppliers (backhauls), while constraint (7) avoids empty running on the way out. Starting service times are calculated in constraints (8) and (9). These constraints also avoid subtours. Time windows are imposed by constraints (10). Constraints (11) avoid exceeding the maximum allowable driving time. Balance of flow is described through constraints (12) and (13). Constraints (14)–(17) are used to restrict the total load a vehicle carries.

The goal of the problem is to construct several routes minimizing the sum of internal and external costs. The internal costs (IC) associated with a given route is composed of five major items: costs of driver (DRC), energy costs (ENC), fixed cost of vehicles–investment, inspection, insurance- (FXC), maintenance costs (MNC) and toll costs (TLC). In addition, the external costs (EC) and social effects of transportation activities are considered. They are composed of: climate change costs (CCC), air pollution costs (APC), noise costs (NSC) and accidents costs (ACC).

$$\begin{aligned} Minimize\ IC + EC = {} & (DRC + ENC + FXC + MNC + TLC) \\ & + (CCC + APC + NSC + ACC) \end{aligned} \tag{18}$$

The mathematical forms of the aforementioned components shown in Eq. (18) are presented below.

$$DRC = \sum_{k=1}^{m} p^k y_0^k \tag{19}$$

$$ENC = \sum_{i=0}^{n} \sum_{\substack{j=0 \\ j \neq i}}^{n} \sum_{k=1}^{m} \sum_{r=1}^{R} fc^r \delta^{kr} d_{ij} (fe^k x_{ij}^k + feu^k f_{ij}^k) \tag{20}$$

$$FXC = \sum_{i=1}^{n} \sum_{k=1}^{m} fx^k x_{0i}^k \tag{21}$$

$$MNC = \sum_{i=0}^{n} \sum_{\substack{j=0 \\ j \neq i}}^{n} \sum_{k=1}^{m} mn^k d_{ij} x_{ij}^k \tag{22}$$

$$TLC = \sum_{i=0}^{n} \sum_{\substack{j=0 \\ j \neq i}}^{n} \sum_{k=1}^{m} tl_{ij} x_{ij}^k \tag{23}$$

$$CCC = \sum_{i=0}^{n} \sum_{\substack{j=0 \\ j \neq i}}^{n} \sum_{k=1}^{m} \sum_{r=1}^{R} pe^{CO2} \delta^{kr} ef^{CO2,r} d_{ij} (fe^k x_{ij}^k + feu^k f_{ij}^k) \tag{24}$$

$$APC = \sum_{i=0}^{n} \sum_{\substack{j=0 \\ j \neq i}}^{n} \sum_{k=1}^{m} \sum_{r=1}^{R} \sum_{t=1}^{T} \sum_{p=1}^{P} pe^p \delta^{kr} \gamma^{kt} ef^{p,t} d_{ij} x_{ij}^k \tag{25}$$

$$NSC = \sum_{i=0}^{n} \sum_{\substack{j=0 \\ j \neq i}}^{n} \sum_{k=1}^{m} ne\ d_{ij} f_{ij}^k \tag{26}$$

$$ACC = \sum_{i=0}^{n} \sum_{\substack{j=0 \\ j \neq i}}^{n} \sum_{k=1}^{m} ae \, d_{ij} f_{ij}^{k} \tag{27}$$

The set of parameters used in the above expressions are:

p^k	pay of driver k per unit time
fc^r	unit cost of fuel type r
fe^k	fuel consumption for the empty vehicle k
feu^k	fuel consumption per unit of additional load in vehicle k
δ^{kr}	equal to 1 if vehicle k uses the fuel type r
fx^k	the fixed cost of vehicle k
mn^k	costs of preventive maintenance, repairs and tires per km of vehicle k
tl_{ij}	costs of tolls associated with arc (i, j)
pe^{CO2}	unit price per ton of CO_2 emitted
$ef^{CO2,r}$	emission factor, amount of CO_2 emitted per unit of fuel r consumed
pe^p	the unit price per ton of the pollutant p emitted
$ef^{p,t}$	amount of pollutant p emitted from technology vehicle t per km travelled
γ^{kt}	equal to 1 if vehicle k belongs to technology t
ne	costs of noise emissions per ton of load carried and per km travelled
ae	costs of accidents per ton of load carried and per km travelled

4.2 A Heuristic Algorithm to Solve the HVRP

Our approach to solving HVRP is based on the savings heuristic originally proposed for the routing problem by (Clarke and Wright 1964). Because Clarke and Wright heuristic was not designed for heterogeneous fleet with capacity constraints, an extension that evaluates the benefit of merging two routes and then assigns a vehicle with feasible capacity to satisfy the demand has been incorporated.

The extended algorithm includes the ability to perform with a heterogeneous fleet. Thus, the route costs are calculated for all vehicles, regardless of meeting with the capacity condition, and then the saving costs for all vehicles can be obtained. Finally, the candidate routes to be fused can be selected based on several criteria, a highest saving or a highest average saving. When two routes are, fused an assignment problem is solved to get the best assignment of vehicles to routes (based on cost criteria).

In the initial iteration process, any generated solution has a high probability to be unfeasible, because the number of routes may exceed the number of available vehicles. Also in the last algorithm iterations, routes cannot be fused because demands exceed the vehicle capacity. Therefore, the heuristic has to select the best solution that will be acceptable compared with the solutions in subsequent iterations.

Finally, it's possible that the problem will be not feasible (for example with high demands in delivery points and few vehicles with low capacity). Given this case, the algorithm has a feasible verification process. The algorithm diagram is represented in Fig. 1

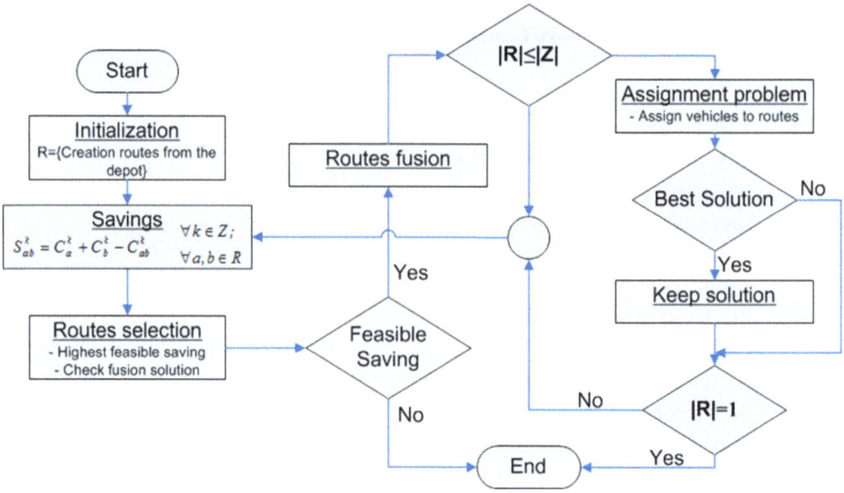

Fig. 1 Savings algorithm diagram

Let Z the set of vehicles, N the set of customers to delivery and R the set of routes. The algorithm starts by forming for each client a route that connects it to the depot, having a set R of N tours. These routes are iteratively joined until the algorithm stops when R has only one route or when unfeasible joining routes occurs. The algorithm keeps the best found solution when the number of vehicles is equal to the number of routes and later.

4.2.1 Savings Calculation

After each fusion route process, the algorithm calculates for each vehicle, the savings cost between each pair of routes from set R (28), obtaining a savings matrix for each vehicle. Vehicle availabilities and capacities conditions are not taken into account in savings calculation.

$$S^k_{ab} = C^k_a + C^k_b - C^k_{ab} \quad \forall k \in Z; \quad \forall a, b \in R \tag{28}$$

where:

S^k_{ab} is the saving between routes a and b when performed with a vehicle k

C^k_{ab} is the total cost of fusion routes a and b when performed with a vehicle k.

Once the savings matrix for each vehicle is obtained, average values (29) are calculated. The highest value of this matrix designates the two candidate routes to be joined.

$$S_{ab} = \frac{\sum_{\forall k} S^k_{ab}}{m} \quad \forall k \in Z; \quad \forall a, b \in R \tag{29}$$

4.2.2 Check Fusion Routes

To guarantee feasibility, a fusion route check procedure is considered. It orders the routes from higher to lower demand and sequentially routes are assigned to the smaller capacity feasible vehicle. If all routes demands are linked into vehicles, the solution is feasible and routes are joined; otherwise the next largest saving value is chosen and the check procedure is repeated.

4.2.3 Keep the Best Solution

Every time the number of routes in R is less or equal to the number of available vehicles, it is possible to obtain a new feasible solution for the problem. For this purpose, the Hungarian Algorithm is used. It is a combinatorial optimization algorithm which solves the assignment problem in polynomial computing time.

When two routes have been joined and the condition to obtain a solution is satisfied, the algorithm evaluates the assignment problem and keeps the solution if it is better than a previous found.

4.2.4 Lambda OPT Framework

Savings Algorithm ends with 2-optimal and 3-optimal procedures applied to each of the found routes from the feasible solution to improve them.

5 A Numerical Example

In this section, we use a four-node illustrative example to show the differences between using three objective functions: minimizing the total distance travelled (1), minimizing the total internal costs (2) and minimizing the total internal and external costs (3). We also study the traditional CVRP with Heterogeneous Fleet (a), versus the effect of adding Backhaul (b), adding also maximum allowable driving Time (c), and adding also Time Window (d). Thus, 12 instances from the HVRPTWB are solved using optimization software.

We consider the four node network of Fig. 2, with 3 different vehicles at node 0 to serve customers 1, 2 and 3. We consider an average speed of 50 km/h on each arc. Then the driving times t_{ij} between nodes are 1, 2 and 2.24 h, depending on the length of the arc. We assume a homogeneous load demanded by each node as $D_i = 8$ ton. Service times are set to $S_i^k = 1$ h in all nodes by all vehicles, and there are no toll costs.

Table 1 shows the parameters associated to each vehicle of the fleet. Table 2 shows the parameters associated to fuel unit costs, external unit costs and emission factors of vehicle types used. As mentioned above, 12 instances are modeled using the MILP problem. In case (b) we consider a backhaul in node 2 with a demand of

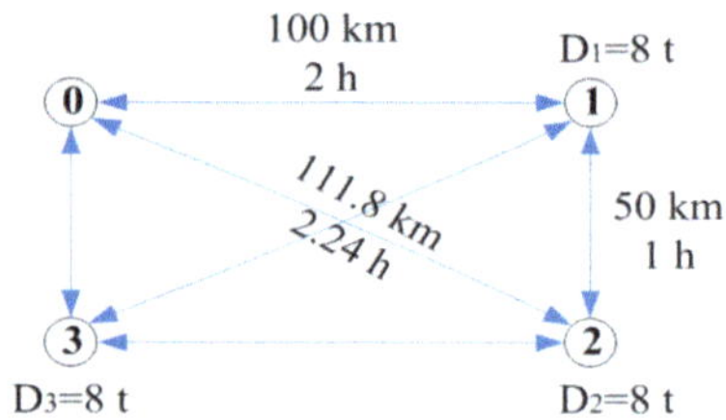

Fig. 2 Four node example

Table 1 Fleet parameters

Vehicles (k)	1	2	3
q^k (tons)	9.5	18	9.5
p^k (€/h)	19.89	21.40	19.89
Type of fuel (r)	Diesel	Diesel	Diesel
fe^k (l/100 km)	17.50	19.80	17.50
feu^k (l/ton·100 km)	1.05	0.75	1.05
fx^k (€/day)	42.65	54.60	42.65
mn^k (€/km)	0.0590	0.0787	0.0590
Technology (t)	Rigid; 12–14_t; Euro_IV t = 2	Rigid; 20–26_t; Euro_IV t = 3	Rigid; 12–14_t; Euro_II t = 1

Table 2 Unit costs

fc^{diesel} (2010€/l)	0.9009		
pe^{CO2} (2010€/ton)	25		
$ef^{CO2,diesel}$ (kg/l)	2.67		
Pollutant (p)	NOx	NMVOC	PM
pe^p (2010€/ton)	3439.8	529.2	76337.1
$ef^{p,1}$ (g/km)	5.50	0.207	0.1040
$ef^{p,2}$ (g/km)	2.65	0.008	0.0161
$ef^{p,3}$ (g/km)	3.83	0.010	0.0239
ne (2010€/ton-km)	0.00648		
ae (2010€/ton-km)	0.00635		

$D_2 = -8$ ton. In case (c) we also assume a maximum driving time for each vehicle of 8 h. And finally in case (d) we also set a time window in node 1 of [3, 5 h]. We have used CPLEX 11.1 with its default settings to solve the 12 MILP instances. Eight different solutions have been found (Table 3). The solutions associated to each instance and the objective functions are illustrated in Table 4.

Some implications of the results presented in Table 4 are as follows.

Optimal solutions which consider the traditional objective function of minimizing total distance travelled (Sol#1 to Sol#4) are not optimal in some cases when the objective function includes costs parameters. But optimal solutions which consider internal and external costs in the objective function (Sol#5, #2, #6 and #8) are also optimal minimizing distances or internal costs. The reason is that minimizing internal costs is quite similar to minimizing distances.

Table 3 Different optimal solutions

Solution	Vehicle	Optimal route	Load (ton)	Arrival time (h)
#1	1	0-3-0	8-0	3
	2	0-2-1-0	16-8-0	7.24
#2	2	0-3-1-2-0	16-8-0-8	9.48
#3	1	0-3-0	8-0	3
	3	0-1-2-0	8-0-8	7.24
#4	1	0-1-0	8-0	6
	2	0-3-2-0	8-0-8	7.24
#5	1	0-3-0	8-0	3
	2	0-1-2-0	16-8-0	7.24
#6	1	0-1-2-0	8-0-8	7.24
	3	0-3-0	8-0	3
#7	1	0-1-0	8-0	6
	3	0-3-2-0	8-0-8	7.24
#8	1	0-3-2-0	8-0-8	7.24
	3	0-1-0	8-0	6

Table 4 Solutions and values of the three objective functions for all the instances

Instance	Solution	Objective function 1 Total distances	Objective function 2 Total inernal costs	Objective function 3 Total costs
1a	#1	361.8[a]	419.5	463.9
1b	#2	323.6[a]	358.3[a]	402.0[a]
1c	#3	361.8[a]	387.2[a]	428.1
1d	#4	461.8[a]	498.6	538.6
2a	#5	361.8[a]	418.2[a]	460.0[a]
2b	#2	323.6[a]	358.3[a]	402.0[a]
2c	#6	361.8[a]	387.2[a]	425.2[a]
2d	#7	461.8[a]	468.6[a]	511.6
3a	#5	361.8[a]	418.2[a]	460.0[a]
3b	#2	323.6[a]	358.3[a]	402.0[a]
3c	#6	361.8[a]	387.2[a]	425.2[a]
3d	#8	461.8[a]	468.6[a]	510.5[a]

[a] Optimal solution with that objective function

When a heterogeneous fleet is considered, adding external costs implies the selection of the less pollutant vehicles or the assignment of longer routes to those vehicles (Sol#7 vs. Sol#8), maintaining minimum total internal costs.

Depending on the type of VRP, the analysis of performance measures must be different. Solutions including backhauls reduce all the costs [see Table 4, Inst. (b) vs. Inst. (a)]. But adding time constraints increase the costs [see Table 4, Inst. (d) or Inst. (c) vs. Inst. (b)]. Using the total costs allows comparing different solutions and selecting the most appropriate. For example, Sol#8 is better than Sol#7 for the external cost, and also Sol#7 is better than Sol#4 for the internal and external costs.

6 A Real Case Application

This section shows the results obtained by analyzing the delivery activity of a Spanish leading supermarket chain in the region of Huelva, a southern Spanish province.

In this region, the network consists of 17 supermarkets or delivery points, which are spread throughout the province (See Fig. 3) served directly from a central distribution center (Depot). Service times are set to $S_i^k = 1$ h in all nodes by all vehicles, and there are no toll costs.

The optimal resolution of the model has been made with CPLEX 11.1 with default parameters in a 3, 30 GHz Intel(R) Core(TM) i5-2400 CPU.

As in the above example, it is used the same heterogeneous fleet to show differences in the use of three different objective functions: minimizing the total distance traveled (1), minimizing internal costs (2) and minimizing internal and external costs (3), taking into account the capacity constraints on each one. The number of vehicles of each class is doubled, to guarantee a feasible solution to the problem.

It is also compared this problem (type *a*), to the inclusion of a maximum driving time of 8 h (type *c*) and limitations of time windows in nodes (type *d*). In problems type *a*, *c* and *d*, a maximum computing time of one, two and four hours has been established respectively and a gap value is obtained. This gap value indicates the percentage of the search space that still remains to be processed with a tolerance of 10^{-4}.

Data concerning the location in geographic coordinates of the distribution and delivery points are summarized in Table 5.

Fig. 3 Distribution and delivery point's locations

Table 5 Distribution and delivery points location

Node	Denomination	Geographic coordinates
0	Depot	37.36777, −6.251307
1	Aljaraque	37.269481, −7.028072
2	Almonte	37.2602, −6.515853
3	Aracena	37.897758, −6.56756
4	Ayamonte	37.22076, −7.399118
5	Bollullos Par del Condado	37.350237, −6.540298
6	Cartaya	37.291339, −7.131093
7	Gibraleón	37.376795, −6.963138
8	Huelva 1	37.262245, −6.959466
9	Huelva 2	37.264112, −6.942893
10	Huelva 3	37.278407, −6.930155
11	Huelva 4	37.25542, −6.932226
12	La Palma del Condado	37.385996, −6.556054
13	Lepe 1	37.258957, −7.196568
14	Lepe 2	37.206551, −7.235068
15	Moguer	37.279747, −6.834899
16	Punta Umbría	37.187369, −6.975337
17	Valverde del Camino	37.578435, −6.752129

The costs of travelling between each two customers, the distances and the travelling time, have been obtained using the Google Maps application.

It is assumed a homogeneous load demanded by each node as $D_i = 1$ ton. In this problem, it has been chosen the same heterogeneous fleet of case 2 (Table 1) but the number of vehicles of each class is doubled to guarantee a solution, being vehicles number 4, 5, 6 the same than 1, 2 and 3 respectively.

Table 6 shows the nodes chosen for this problem with their demands and time windows, while Tables 7 and 8 show the results obtained.

As Table 8 shows, the solutions to the problems, do not ensure that they are optimal, since the gap value obtained indicates that the search space has not been totally explored. This can be tested in the solutions found for every type of problem of minimizing total costs, which have smaller internal costs than the solution found to minimizing those costs.

These results suggest that heuristics approaches are necessary to find high quality solutions to the environmental model.

7 Experimental Results

To analyze the behavior of the heuristics described above, it has been programmed in C++. Computational tests have been performed on the real case of study described in Sect. 6, and also on eight HVRP large instances, well-known as Taillard problems.

Table 6 Demands and time windows for the 17 nodes problem

Node	Demand (ton)	Time window (h)
0	–	[0–8]
1	1	[4–8]
2	1	[4–8]
3	1	[0–4]
4	1	[0–4]
5	1	[4–8]
6	1	[0–4]
7	1	[0–4]
8	1	[0–4]
9	1	[0–4]
10	1	[0–4]
11	1	[4–8]
12	1	[4–8]
13	1	[4–8]
14	1	[0–4]
15	1	[0–4]
16	1	[0–4]
17	1	[0–4]

In the real case application, the *type a* problem (without time window constraints) has been solved using the savings heuristic proposed in Sect. 4. The solutions obtained for the three different objective functions (minimizing the total distance traveled, minimizing internal costs and minimizing internal and external costs) are compared respect to the best solution value obtained with CPLEX in the established computing time.

In order to validate the heuristic behavior for large problems, a set of 8 instances from the HVRP literature have been solved and compared respect to the best known solutions for each instance.

7.1 The Real Case Application

Type *a* problem does not have time windows (TW) constraints. Table 9 shows the results obtained and compare the heuristic and the better solution found with the optimization software.

With respect to solution quality, the heuristic savings algorithm obtained the same solution as CPLEX for minimizing the distance traveled. The deviation from the best solution found by CPLEX in the established computing time was quite small (less than 1 %).

Table 7 Route and travel time solutions for the 17 nodes problem

Problem	Objetive function	Vehicle	Route	Time (h)
Type a (no. max. driving time)	Distance	3	0-5-0	1.60
		5	0-12-17-3-7-4-14-13-6-1-16-8-9-11-10-15-2-0	23.86
	Internal costs	1	0-5-2-12-17-3-0	8.96
		2	0-15-7-10-11-9-8-16-1-6-13-14-4-0	16.61
	Internal + external costs	1	0-9-8-11-10-7-15-17-3-0	12.99
		4	0-5-2-4-14-13-6-1-16-12-0	13.43
Type c (max. driving time 8 hours)	Distance	2	0-5-12-17-3-0	7.44
		4	0-16-1-8-9-11-0	7.55
		5	0-15-10-7-2-0	6.45
		6	0-13-6-14-4-0	7.14
	Internal costs	1	0-6-13-4-14-0	7.01
		3	0-5-2-9-11-15-0	7.47
		5	0-7-1-16-8-10-0	7.57
		6	0-12-17-3-0	6.39
	Internal + external costs	1	0-6-13-14-4-0	6.84
		4	0-5-12-17-3-0	7.44
		5	0-15-7-16-1-0	6.79
		6	0-10-11-9-8-2-0	7.31
Type d (time windows)	Distance	1	0-7-10-15-2-0	6.48
		3	0-3-17-12-5-0	7.44
		5	0-9-8-16-1-11-0	7.59
		6	0-6-4-14-13-0	7.00
	Internal costs	1	0-15-10-7-12-0	6.22
		3	0-4-14-6-11-0	7.26
		4	0-3-17-5-2-0	7.95
		6	0-9-8-16-1-13-0	7.84
	Internal + external costs	1	0-9-16-10-12-2-0	7.94
		3	0-6-4-14-13-0	7.00
		4	0-15-7-8-11-1-0	7.74
		6	0-3-17-5-0	6.42

7.2 Taillard Problems

In this section we describe the performance of the heuristic algorithm on the set of 8 instances from Taillard problems (13–20) and compare the solutions to the results reported in the literature. Taillard problems do not have TW constraints.

As illustrated in Table 10, the heuristic savings algorithm seems to perform well, with an average percentage difference from the best known solution of 11.09 %. It is important to mention that the proposed heuristic has been developed to solve the HVRP with internal and external costs, and Taillard instances are prepared to solve the HVRP with Dependent routing costs.

Table 8 Solution values of the three objective functions for the 17 nodes problem

Problem	Objetive function	Distance (km)	Internal costs (€)	External costs (€)	Total costs (€)	Gap (%)	Run time (s)
Type a	Distance	520.80	801.97	75.55	877.52	8.61	3,600
	Internal costs	567.60	779.88	39.13	819.01	74.39	3,600
	Internal + external costs	628.60	768.51	44.65	813.15	70.91	3,600
Type c	Distance	859.10	1.000.41	47.59	1.048.01	45.70	7,200
	Internal costs	851.20	965.12	47.41	1.012.53	77.46	7,200
	Internal + external costs	852.20	963.09	42.21	1.005.30	73.82	7,200
Type d	Distance	849.40	970.32	55.15	1.025.47	29.70	14,400
	Internal costs	903.70	968.57	56.46	1.025.03	61.11	14,400
	Internal + external costs	888.40	960.91	54.86	1.015.77	62.31	14,400

Table 9 Comparison on the algorithm on type a problem

Problem	Obj func.pollution-routing problem	CPLEX (max 3,600 s)			Savings algorithm	
		Cost (€)	Time (s)	Solut. (%)	Cost (€)	Error (%)
17a	Distance	520.8	3,600	8.61	520.80	0.00
	Internal costs	779.88	3,600	74.39	786.78	0.88
	Internal + external costs	813.15	3,600	70.91	822.03	1.09

Table 10 Comparison on the algorithm on Taillard problems

Problem	Nodes	Best known solution	Savings algorithm	
			Cost (€)	Error (%)
13	50	1,517.84	1,720.57	13.36
14	50	607.53	661.02	8.80
15	50	1,015.29	1,110.21	9.35
16	50	1,144.94	1,248.78	9.07
17	75	1,061.96	1,159.59	9.19
18	75	1,823.58	1,990.70	9.16
19	100	1,117.51	1,327.57	18.80
20	100	1,534.17	1,702.80	10.99

8 Conclusions

Research in Vehicle Routing Problems still needs more extensive study in the areas of environmental issues and heterogeneous fleet with time windows constraints. There are few publications that address eco-efficiency objectives in vehicle routing and scheduling problems. The need for companies to incorporate these external factors as part of their planning and operational process is forcing traditional VRP studies to model the fuel consumption, the pollutants emissions and other external impacts within the objective function.

In this chapter, external factors were incorporated in a mixed integer linear programming model for solving the traditional VRP with realistic assumptions such as a limited number of heterogeneous vehicles, time windows constraints and backhauls (HVRPTWB). Before the model was suggested, literature reviews on VRP with heterogeneous fleet and on VRP and sustainability were presented. Due to the intrinsic difficulty of this type of routing problems, solution approaches in the literature are heuristic algorithms. In this chapter, a new heuristic algorithm that extends the savings algorithm of Clark and Wright was developed to solve the HVRP.

A numerical example was showing to illustrate and validate the model. A real case application for the HVRPTW was presented and solved using optimization software with a limited computing time and using the proposed heuristic in some cases. Finally the heuristic was tested on the set of 8 instances from Taillard for HVRPD. Computational results show good quality solutions and may be used to solve the new eco-efficiency model without time windows.

Further research may lead to the development of a new metaheuristic algorithm to incorporate time windows. Also further work on quantifying the congestion factor is in order. Finally, we are incorporating variations in speed as part of the model.

Acknowledgments This research has been fully funded by the Spanish Ministry of Science and Innovation and FEDER through grants DPI2008-04788 and by the Andalusia Government through grants P10-TEP-6332.

References

Baldacci R, Mingozzi A (2009) A unified exact method for solving different classes of vehicle routing problems Math. Program 120:347–380

Baldacci R, Battarra M, Vigo D (2008) Routing a heterogeneous fleet of vehicles. In: The vehicle routing problem: latest advances and new challenges. Springer, Berlin, pp 11–35

Balinski M, Quandt R (1964) On an integer program for a delivery problem. Oper Res 12:300–304

Bektas T, Laporte G (2011) The pollution-routing problem. Transp Res Part B: methodological 45(8):1232–1250. doi:10.1016/j.trb.2011.02.004

Bickel P, Friedrich R, Link H, Stewart L, Nash C (2006) Introducing environmental externalities into transport pricing: measurement and implications. Transp Rev 26(4):389–415

Brandão J (2009) A deterministic tabu search algorithm for the fleet size and mix vehicle routing problem. Eur J Oper Res 195:716–728

Bräysy O, Dullaert W, Hasle G, Mester D, Gendreau M (2008) An effective multirestart deterministic annealing methauristic for the fleet size and mix vehicle routing problem with time windows. Transp Sci 42(3):371–386

Choi E, Tcha D (2007) A column generation approach to the heterogeneous fleet vehicle routing problem. Comput Oper Res 34(7):2080–2095

Clarke G, Wright J (1964) Scheduling of vehicles from a central depot to a number of delivery points. Oper Res 12(4):568–581
Cordeau J-F, Laporte G, Savelsbergh MWP, Vigo D (2007) Vehicle routing. In: Barnhart C, Laporte G (eds) Transportation, handbooks in operations research and management science, vol 14. Elsevier, Amsterdam, pp 367–428
Dantzig GB, Ramser JH (1959) The truck dispatching problem. Manage Sci 6(1):80–91
Dell'Amico M, Monaci M, Pagani C, Vigo D (2006) Heuristic approaches for the fleet size and mix vehicle routing problem with time windows. Technical report, DISMI, University of Modena and Reggio Emilia, Italy
Dullaert W, Janssens GK, Sörensen K, Vernimmen B (2002) New heuristics for the fleet size and mix vehicle routing problem with time windows. J Oper Res Soc 53(11):1232–1238
Dyckhoff H, Lackes R, Reese J (2004) Supply chain management and reverse logistics. Springer, Berlin
EMEP/EEA (2010) Emission inventory guidebook 2009, updated June 2010. European Environment Agency (EEA)
Erdogan S, Miller-Hooks E (2012) A green vehicle routing problem. Transp Res Part E 48:100–114
Figliozzi M (2010) Vehicle routing problem for emissions minimization. Transp Res Rec: J Transp Res Board 2197:1-7. Transportation Research Board of the National Academies, Washington, D.C. doi: 10.3141/2197-01
Gendreau M, Laporte G, Musaraganyi C, Taillard ED (1999) A tabu search heuristic for the heterogeneous fleet vehicle routing problem. Comput Oper Res 26(12):1153–1173
Gheysens FG, Golden BL, Assad AA (1984) A comparison of techniques for solving the fleet size and mix vehicle routing problem. OR Spectr 6(4):207–216
Golden B, Assad A, Levy L, Gheysens FG (1984) The fleet size and mix vehicle routing problem. Comput Oper Res 11:49–66
Hoff A, Andersson H, Christiansen M, Hasle G, Løkketangen A (2010) Industrial aspects and literature survey: fleet composition and routing. Comput Oper Res 37(12):2041–2061
Imran A, Salhi S, Wassan NA (2009) A variable neighborhood-based heuristic for the heterogeneous fleet vehicle routing problem. Eur J Oper Res 197:509–518
INFRAS/IWW (2004) External costs of transport: update study, INFRAS, Zurich
INFRAS, CE Delft, ISI and University of Gdansk (2008) Handbook on estimation of external cost in the transport sector. Produced within the study: internalisation measures and policies for all external cost of transport (IMPACT), Delft, CE
Jabali O, Van Woensel T, de Kok AG (2009) Analysis of travel times and CO2 emissions in time-dependent vehicle routing. Technical report Eindhoven University of Technology
Kara I, Kara BY, Yetis MK (2007) Energy minimizing vehicle routing problem. In: Dress A, Xu Y, Zhu B (eds) Combinatorial Optimization and Applications, Lecture Notes in Computer Science, vol 4616. Springer, Berlin/Heidelberg, pp 62–71
Lee Y, Kim J, Kang K, Kim K (2008) A heuristic for vehicle fleet mix problem using tabu search and set partitioning. J Oper Res Soc 59:833–841
Li F, Golden BL, Wasil EA (2006) A record-to-record travel algorithm for solving the heterogeneous fleet vehicle routing problem. Comp Oper Res 34(9):2533−2864
Lima CMRR, Goldbarg MC, Goldbarg EFG (2004) A memetic algorithm for the heterogeneous fleet vehicle routing problem. Electron Notes Discrete Math 18:171–176
Liu FH, Shen SY (1999) The fleet size and mix vehicle routing problem with time windows. J Oper Res Soc 50(7):721–732
Liu S, Huang W, Ma H (2009) An effective genetic algorithm for the fleet size and mix vehicle routing problems. Transp Res Part B Methodol 45:434–445
Maden W, Eglese RW, Black D (2010) Vehicle routing and scheduling with time varying data: a case study. J Oper Res Soc 61(3):515–522
Miller CE, Tucker AW, Zemlin RA (1960) Integer programming formulation of traveling salesman problems. J ACM 7(4):326–329

Ochi LS, Vianna DS, Drummond LMA, Victor AO (1998) A parallel evolutionary algorithm for the vehicle routing problem with heterogeneous fleet. Parallel Distrib Process 1388:216–224

Palmer A (2007) The Development of an integrated routing and carbon dioxide emissions model for goods vehicles. Ph.D. thesis, Cranfield University, School of Management

Penna P, Subramanian A, Ochi L (2011) An iterated local search heuristic for the heterogeneous fleet vehicle routing problem. J Heuristics. doi:10.1007/s10732-011-9186-y

Pessoa A, Uchoa E, de Aragão MP (2009) A robust branch-cut-and-price algorithm for the heterogeneous fleet vehicle routing problem. Network 54(4):167–177

Prins C (2002) Efficient heuristics for the heterogeneous fleet multitrip VRP with application to a large-scale real case. J Math Model Algorithms 1:135–150

Prins C (2009) A GRASP evolutionary local search hybrid for the vehicle routing problem. In: Bio-inspired algorithms for the vehicle routing problem, studies in computational intelligence, vol 161. Springer, Berlin, pp 35–53

Renaud J, Boctor FF (2002) A sweep-based algorithm for the fleet size and mix vehicle routing problem. Eur J Oper Res 140(3):618–628

Salhi S, Rand GK (1993) Incorporating vehicle routing into the vehicle fleet composition problem. Eur J Oper Res 66:313–330

Sbihi A, Eglese RW (2007) The relationship between vehicle routing and scheduling and green logistics—a literature survey. Management school working paper, 2007/027, Lancaster University, Lancaster

Sniezek J, Bodin L (2002) Cost models for vehicle routing problems. In: Proceedings of the 35th Hawaii international conference on system sciences. pp 1403–1414

Solomon MM (1987) Algorithms for the vehicle routing and scheduling problems with time window constraints. Oper Res 35(2):254–265

Taillard ED (1999) A heuristic column generation method for the heterogeneous fleet VRP. RAIRO Recherche Opérationnelle 33(1):1–14

Tarantilis CD, Kiranoudis CT, Vassiliadis VS (2003) A list based threshold accepting metaheuristic for the heterogeneous fixed fleet vehicle routing problem. J Oper Res Soc 54(1):65–71

Tarantilis CD, Kiranoudis CT, Vassiliadis VS (2004) A threshold accepting metaheuristic for the heterogeneous fixed fleet vehicle routing problem. Eur J Oper Res 152(1):148–158

Toth P, Vigo D (2002) The vehicle routing problem. SIAM, SIAM monographs on discrete mathematics and applications, pp 195–224

Ubeda S, Arcelus FJ, Faulin J (2011) Green logistics at Eroski: a case study. Int J Prod Econ 131:44–51

Xiao Y, Zhao Q, Kaku I, Xu Y (2012) Development of a fuel consumption optimization model for the capacitated vehicle routing problem. Comput and Oper Res 39(7):1419–1431

Yaman HD (2006) Formulations and valid inequalities for the heterogeneous vehicle routing problem. Math Program 106(2):365–390

Assessment of Criticality of Spare Parts Using the Method of Multi-Criteria Decision Making

Karolina Kolinska and Dawid Dolinski

Abstract Useful element in the process of inventory control of spare parts is the information if spare parts is critical from the standpoint of the production process, so whether it is necessary to keep it in stock. This decision can be made by using the method of multicriteria decision making Analytic Hierarchy Process (AHP). An important element in the application of this method in enterprises is to engage maintenance and the Purchasing Department when defining the criteria that determine the criticality of spare parts. This chapter describes an example of using AHP method in the assessment of criticality of spare parts, which is the original result of the work of the authors.

Keywords AHP method · Spare parts

1 The Essence of Spare Parts Management

Most of production companies have in its organizational structure Maintenance Department (part of the enterprises have Maintenance Department as outsourcing), which is responsible for the maintenance of production in continuous operation. The function of spare parts inventories is to assist the maintenance operations in keeping machinery and the devices included in the production lines in operating condition.

K. Kolinska (✉) · D. Dolinski
Institute Logistics and Warehousing, Poznan, Poland
e-mail: karolina.kolinska@ilim.poznan.pl

D. Dolinski
e-mail: dawid.dolinski@ilim.poznan.pl

M. G. Erechtchoukova et al. (eds.), *Sustainability Appraisal: Quantitative Methods and Mathematical Techniques for Environmental Performance Evaluation*, EcoProduction, DOI: 10.1007/978-3-642-32081-1_11,

An important element is apprioprate definition of spare parts. According to the standards (norm) DIN 24420, spare parts are items (also called parts), groups of elements (named also the modules, and submodules) or complete products that are used to replace the damaged, wasted or defective parts, modules, or products. APICS standards define spare parts as modules, components, and elements that are planned to be used without modification to replace an original part (APICS 2004). You can define spare parts as follow (Biedermann 2008):

- Spare parts—components, groups or complete products, except for damaged, worn or missing parts or groups of products,
- Replaceable spare parts—spare parts which are assigned to one or several machines and are used in case of failure of machinery or equipment,
- Parts ready to be replaced, which can be use due to its design destination and usually not regenerated because of low economic value,
- Small parts—generally used spare parts, mainly the standard and low values.

Nowadays, increasing attention is paid to problems of ecological management of production processes in enterprises and supply chains. One of the effective means of ecological production management is to maintain the operational readiness of available machinery in the company, which determines not only to obtain the planned production capacity at a reasonable or acceptable level of production costs but also contribute to the minimization waste of natural resources. The increased interest in maintenance is not only accelerated by the need to start the modernization processes and replacement of machinery and technological equipment, but to a greater extent due to the progressive specialization of production and the need for rapid deployment of new products for the consumer market. In the literature review can be found that effective maintenance organization is the cheapest and most effective means of improving the operating efficiency of fixed assets and has a direct impact on the proecological strategy of the company (Znoj and Zawadzka 1985).

One of the important function that implement Maintenance Department is optimizing the availability of equipment at minimum cost. Other functions carried out by Maintenance Department are (Mikler 2008):

- Safety of humans and the environment,
- Efficiency of production,
- Risk management,
- Efficient energy consumption,
- Quality of products and services offered.

The modern approach to Maintenance Department is mainly characterized by (Legutko 2009):

- Avoid, reduce or eliminate defects and not only prevention,
- Carry out about safety of people and the environment, product quality and level of customer service, and not only the costs optimizing,

- Rejection the idea of binding the age of equipment and machinery from the intensity of their damage,
- Specifying the frequency of maintenance activities on the basis of the symptoms of damage and not on the basis of indicators of failure of equipment and machinery,
- Development of common methods of maintenance for identical machines, which have the same rules for the operation, functions and the expected standards of implementation,
- Independent develop programs of maintenance in the companies, reasonably comply the recommendations of the supplier, and not treating supplier's rules as the only authorized for their development,
- Development of programs of maintenance by service maintenance and operators,
- Approve of the involvement of employees from all management levels of the company, it means that operation maintenance as a key factor of success of the companies,
- Arrangement of Maintenance Department as a strategic area of the company, and not only as a secondary department.

For successful execution of the maintenance process it is necessary to reduce production costs, increasing the reliability of machinery, but also increase the efficiency of the production process. Extremely important issue of maintenance is effective control of the inventory of spare parts which have a direct impact on the continuous operations of the production process.

One of the most important aspects to control the stock of spare parts is the classification of the spare parts for determining the criticality of these parts from the point of view of the continuity manufacturing process.

In order to manage such a classification, the authors propose using one of the multi criteria methods of decision-making—AHP (Analytic Hierarchy Process).

2 Assessment of Criticality of Spare Parts Using Analytic Hierarchy Process Method

Assessment of criticality of spare parts can be solved in two ways. First, are based only on the experience of employees of Maintenance Department who decide the criticality of the particular spare part. In such a situation, it is difficult to carry out a verification of the results and discuss with the results obtained, as it does not rely on calculations.

To make a good decision we need to know the problem, the need and purpose of the decision, the criteria of the decision, their subcriteria, stakeholders and groups affected and the alternative actions to take. We then try to determine the best alternative, or in the case of resource allocation, we need priorities for the alternatives to allocate their appropriate share of the resources (Saaty 2008). Such opportunities to evaluate the criticality of spare parts provide AHP.

The analytic hierarchy process (AHP) is a mathematical method for analyzing complex decision problems under multiple criteria (Saaty 1995). The management options for a particular decision problem are characterized by their attributes with respect to a set of detailed criteria (Qureshi and Harrison 2003). One of the main advantages of the application of the AHP in practice is the possibility of making an objective choice based on the hierarchy of criteria analyzing. In accordance with the assumptions of the AHP, hierarchy of validity has a structure that involves peeling to the decision-making process in the initial stage, then on establishing evaluation criteria and variants of solutions (Bruno et al. 2012).

Conducting a structured decision-making process is possible if the action plan is established. It is recommended that this process has continued by the following steps (Saaty 2008):

- Define the problem and determine the kind of knowledge sought,
- Structure the decision hierarchy from the top with the goal of the decision, then the objectives from a broad perspective, through the intermediate levels (criteria on which subsequent elements depend) to the lowest level (which usually is a set of the alternatives),
- Construct a set of pairwise comparison matrices. Each element in an upper level is used to compare the elements in the level immediately below with respect to it,
- Use the priorities obtained from the comparisons to weigh the priorities in the level immediately below. Do this for every element. Then for each element in the level below add its weighed values and obtain its overall or global priority. Continue this process of weighing and adding until the final priorities of the alternatives in the bottom most level are obtained.

3 Case Study

Maintain continuity of production by Maintenance Department requires possession of suitable spare parts in stock so that you can quickly run the machinery. There are groups of spare parts, which are critical from the standpoint of the company and they should necessarily be kept in a wide variety of locations and in significant quantities. Some spare parts are not critical, but it is also important to keep them in stock. Others parts and materials are high available on the market and therefore there is no need to keep them in stock. Decision which parts should be allocated to the Group of "critical", "important" or "other" may be carried out using the multi criteria decision-making, in view of the large number of criteria that affect the criticality of spare parts.

As already mentioned multi criteria decision making can be carried out using different methods. In this chapter the authors decided to present methods of AHP on the practical example.

During the implementation of the process of assessment the criticality of spare parts using the method of AHP the following tasks should be performed[1] :

1. *Define the problem*

   Solution: Assessment of criticality of spare parts.

2. *Defining groups of the classification of spare parts—specify which group should divide the spare parts and to define the characteristics of these groups*

   Solution: Spare parts will be divided into three groups:

- Critical—high risk of losing high-margin, long lead times, the small number of potential suppliers, the low index of use—spare parts, which are crucial for the continuity of production and reduction of losses; these parts are not entirely available from suppliers, and in specific cases only available on special order,
- Important—medium risk of losing high-margin, shorter lead times, at least two potential suppliers—spare parts, which are important for the continuity of production; it is possible to stop production line, because it does not generate huge losses,
- Other—small risk of losing high-margin, short lead times, large number of potential suppliers, a high index of use—spare parts that do not cause large losses and are fully available from suppliers.

3. *Defining criteria for the classification of spare parts—create list of the criteria* (7 ± 2), *which are the most relevant from the point of view of assessment the criticality of spare parts.*

Solution: The list of examples, which gives information to take a decision about the criticality of spare part:

- Lead time—delivery time of the material,
- Price—the value of the purchase of one unit of the material (spare part),
- Lost margin—impact on operational risk, the value of the lost margin during the hours/days stationary installations,
- Usage rate—the amount of material consumption in a given period of time in relation to the frequency of releases, how often a spare part was used by the Maintenance Department in the analysed period. The rules are described in Table 4.
- Number of potential suppliers—one supplier/multiple suppliers.

4. *Ranking criteria for the classification of spare parts from the most important to least important—after defining the criteria for the classification of spare parts they should be ranked in terms of their validity (from most to least important).*

[1] Tasks are defined in accordance with the methodology of AHP, but nevertheless they have been tailored to the specific of spare parts.

Table 1 The values used to assess the validity of the individual criteria between each other

Value	Description of relation	Value	Description of relation
1	Both criteria of equal importance	0	Lack of criterium for the material (spare part)
3	Left weakly more important than top	1/3	Top weakly more important than left
5	Left moderately more important than top	1/5	Top moderately more important than left
7	Left strongly more important than top	1/7	Top strongly more important than left
9	Left absolutely more important than top	1/9	Top absolutely more important than left
2, 4, 6, 8	Intermediate values		

Source own study

Solution: The list of criteria, in order from most to least important:

- Lost margin,
- Lead time,
- Price,
- Number of potential suppliers,
- Usage rate

5. *Definition of validity of individual criteria between the criteria (compare pairs)—under this task, you create a matrix containing all defined criteria,*

Then you must compare each criterion (the comparison of pairs) and assign them to the appropriate values. Use the values of individual variables, which are presented in Table 1.

The next step is to determine the normalized weight (normalized eigenvector) for particular criteria—determination of normalized weights (eigenvectors) are calculated as the double designation of products matrix.

Solution:

Table 2 shows the results of the assessment of the validity of the individual criteria between each other.

6. *Collect actual data for the designated classification criteria—for each item or material should be collected actual data for each of the defined criteria. It is important that data relate to the same period and are expressed in the same unit of measure. You should also remember that the actual data relate to the current situation in the company.*

Solution:

Table 3 shows actual data for the individual parts, which have been collected from the company IT system, and also from the Maintenance Department workers.

Table 2 The validity of individual criteria between each other—matrix

Matrix	Analyzed criterion					Normalized weights
	Lost margin (PLN)	Lead time (days)	Price (PLN)	Number of potential suppliers	Usage rate	
Lost margin (PLN)	1.00	3.00	5.00	7.00	8.00	0.5146
Lead time (days)	0.33	1.00	2.00	5.00	7.00	0.2461
Price(PLN)	0.20	0.50	1.00	2.00	5.00	0.1290
Number of potential suppliers	0.14	0.20	0.50	1.00	4.00	0.0762
Usage rate	0.13	0.14	0.20	0.25	1.00	0.0341
Total						1.00

Source own study

Table 3 Actual data for the individual parts

Num..	Item	Data				
		Lost margin (PLN)	Lead time (days)	Price (PLN)	Number of potential suppliers	Usage Rate (–)[a]
1	A	7,000	15	340	0	1
2	B	12,000	18	120	1	2
3	C	130,000	76	2,650	1	3
4	D	100,000	90	8,600	1	4
5	E	153,000	120	3,070	1	4
6	F	12,000	67	1,970	1	4
7	G	78,000	54	56	0	3
8	H	78,000	10	143,000	1	3
9	I	7,000	23	9,300	1	4
10	J	153,000	38	4,600	0	2

Source own study

[a] Usage rate has no measure, each item can be assigned to one of the four groups. The rules are described in Table 4

7. *Designation of compartments (subcriterion) for each of the criteria.*

   Solution:

   Table 4 shows defined value ranges for each criteria. Intervals were determined on the basis of actual data collected for individual parts and taking into account the specificities of the company.

8. *The importance of individual compartments in a given criterion (the comparison of pairs)—In this task, create a matrix for each of the defined criteria (number of the matrix equal to the quantity of the criteria).*

Each of the matrix should consist of the ranges specified for individual criteria (paragraph 6). Then you should compare every compartment of each (the

Table 4 Compartments (subcriterion) for each of the criteria

Main criteria	K1	K2	K3	K4
Lost margin (PLN)	More than 120,000	50,000–120,000	10,000–50,000	Less than 10,000
Lead time (days)	More than 70	40–70	20–40	Less than 20
Price (PLN)	More than 5,000	1,500–5,000	500–1,500	Less than 500
Number of potential suppliers	One supplier (0)	More than one supplier (1)	–	–
Usage rate	Min. 2 issues in the six month (1)	6 month with at least two issues in 3 years (2)	1 issue in 3 years (3)	No issues in 3 years (4)

Source own study

Table 5 The importance of individual compartments for lost margin—matrix

Lost margin (PLN)	K1	K2	K3	K4	Weights
K1	1.00	4.00	6.00	8.00	0.6136
K2	0.25	1.00	3.00	6.00	0.2361
K3	0.17	0.33	1.00	3.00	0.1026
K4	0.13	0.17	0.33	1.00	0.0477
Total					1.00

Source own study

Table 6 The importance of individual compartments for lead time—matrix

Lead time (days)	K1	K2	K3	K4	Weights
K1	1.00	4.00	6.00	8.00	0.6223
K2	0.25	1.00	2.00	6.00	0.2206
K3	0.17	0.50	1.00	2.00	0.1035
K4	0.13	0.17	0.50	1.00	0.0536
Total					1.00

Source own study

comparison of pairs) in the framework of the criterion and assign them to the appropriate values. Use the values of individual dependencies, which are shown in Table 1.

Then designate normalized weights for each of the criteria—determination of normalized weights (eigenvectors) is calculated by the double multipled matrix.

Solution:

Table 5 shows comparison of intervals for the criterion Lost Margin.

Table 6 shows comparison of intervals for the criterion Lead Time.

Table 7 shows comparison of intervals for the criterion of price.

Table 7 The importance of individual compartments for price—matrix

Price (PLN)	K1	K2	K3	K4	Weights
K1	1.00	2.00	6.00	8.00	0.56
K2	0.50	1.00	2.00	4.00	0.25
K3	0.17	0.50	1.00	4.00	0.14
K4	0.13	0.25	0.25	1.00	0.05
Total					1.00

Source own study

Table 8 The importance of individual compartments for Number of potential suppliers—matrix

Number of potential suppliers	K1	K2	Weights
K1	1.00	6.00	0.86
K2	0.17	1.00	0.14
Total			1.00

Source own study

Table 9 The importance of individual compartments for usage rate—matrix

Usage rate	K1	K2	K3	K4	Weights
K1	1.00	3.00	5.00	8.00	0.55
K2	0.33	1.00	3.00	8.00	0.27
K3	0.20	0.33	1.00	8.00	0.15
K4	0.13	0.13	0.13	1.00	0.04
Total					1.00

Source own study

Table 8 shows comparison of intervals for the criterion Number of Potential Suppliers.

Table 9 shows comparison of intervals for the criterion Usage Rate.

9. *Determination of complex weights for each criterion—to calculate the weights for each of the criteria the multiple of a weight-normalized the criterion and the weight of the standard compartment must be determined.*

Solution:

Table 10 shows designated weight made for each criteria and the individual compartments within a given criterion. The weights are necessary to designate the ranges, thanks to which it will be possible to assign individual spare parts to the defined Group (Critical, Important, Other).

10. *Designation of range for different groups—on the basis of the assigned complex, weights must specify ranges groups of the classification of spare parts.*

Table 11 shows fixed value ranges allocation of individual spare parts to the defined Group.

Table 10 The weight of the complex for the compartments within a given criterion

Main criteria	Number of range			
	K1	K2	K3	K4
Lost margin (PLN)	0.316	0.122	0.053	0.025
Lead time (days)	0.153	0.054	0.025	0.013
Price (PLN)	0.072	0.032	0.018	0.007
Number of potential suppliers	0.065	0.011	–	–
Usage rate	0.019	0.009	0.005	0.001

Source own study

Table 11 The value of ranges to allocate of spare parts to the groups

Group	Ranges			
Critical			More than	0.443
Important	From	0.199	To	0.443
Other			Less than	0.199

Source own study

Table 12 The list of spare parts from the assigned groups

Spare part	Total weigh	Group
E	0.560	Critical
C	0.550	Critical
J	0.418	Important
D	0.366	Important
G	0.257	Important
H	0.216	Important
F	0.198	Other
B	0.143	Other
I	0.141	Other
A	0.111	Other

Source own study

11. *Assign each item to the defined groups assortment of spare parts classification*—assignment of individual items of material to the groups of the classification should be calculated on the basis of actual data, designated complex matrix and specific compartments for the various groups of the classification (Table 12).
12. *Develop a final list of items or material with the prescribed classification groups*—analysis of the list of items or material with assigned groups, in particular those that are at the intersection of each group. Then leave the position of material in the proposed group, or transfer it to another group. The final effect of the work will be the material list with assigned classification of groups.

4 Conclusions and Further Research

An assessment of the criticality of spare parts is a long-term process. In particular, when the list of parts covered by the analysis is large, and if the defined evaluation criteria require gathering of data by employees of the Maintenance Department for individual materials, especially when it is not possible to gather data from company IT system. Very laborious is the verification of the results of the classification and any possible changes to be later on discussed in the top management of the company.

Implementation of AHP method in the assessment of criticality of spare parts has the following advantages:

- An organized process of project evaluation,
- Each spare part evaluated within the same criteria,
- Mechanics know and understand the definitions of individual criteria,
- The decision to set criteria and their definitions made by members of the project,
- The impact of a particular criterion on criticality of spare parts is determined by the project team.

However, the assessment of the criticality of spare parts is not the last step in the process of improving the control/inventory management of spare parts. After develop the list of spare parts from the assigned groups classification it is recommended to fulfill the following tasks:

1. *Developing procedures for assessing the criticality of spare parts. During the preparation of procedures very important are*:
   - Repeating the assessment of the criticality of spare parts (frequency),
   - Methods of assessment of the criticality of new spare parts, for which we don't have the history (like warehouse issues, Lead Time, etc.)
   - The place for information in company's IT System, about the group to which you assigned the spare part of the assessment of the criticality of spare parts,
   - Methodology of the inventory management of spare parts are assigned to each group.
2. *Approval of the procedures by the Board of Directors of the company.*
3. *Presentation of the results of the assessment of the criticality of spare parts to the Top Management and the Board, and final approval of the results.*

Full completion of the evaluation process of the criticality of spare parts, requires the involvement of many people in the company, which is laborious and time consuming. Therefore it is recommended to establish the project manager, who will supervise the work, initiate and manage the project schedule.

References

APICS Dictionary (2004) 11th Edition American Production and Inventory Control Society Inc. Falls Church

Biedermann H (2008) Ersatzteil—management. Effiziente Ersatzteillogistik fur Industrieunternehmen. Springer, Berlin, pp 3–9

Bruno G, Esposito E, Genovese A, Passaro R (2012) AHP-based approaches for supplier evaluation: problems and perspectives. J Purchasing and Supply Manage, In Press, Corrected Proof, Available online 20 June 2012, pp 159–172

Legutko S (2009) Development trends in machines operation maintenance. Maintenance Reliab 2:8–16 (in Polish)

Mikler J (2008) Effective management of the maintenance process—a review of methods. Maintenance 10:42–48 (in Polish)

Qureshi ME, Harrison SR (2003) Application of the analytic hierarchy process to riparian revegetation policy options. Small-scale For Econ Manage Policy 2(3):441–458

Saaty TL (2008) Decision making with the analytic hierarchy process. Int J Serv Sci 1(1):83–98

Saaty TL (1995) Decision making for leaders: the analytic hierarchy process for decisions in a complex world. RWS Publications, Pittsburgh, pp 14–27

Znoj T, Zawadzka E (1985) Motivation systems of maintenance services. Inst Eng Ind Organ, Warsaw (in Polish)